Hans Schippers
Westerweel Group: Non-Conformist Resistance Against Nazi Germany

New Perspectives on Modern Jewish History

Edited by
Cornelia Wilhelm

Volume 11

Hans Schippers

Westerweel Group: Non-Conformist Resistance Against Nazi Germany

—

A Joint Rescue Effort of Dutch Idealists and Dutch-German Zionists

This book is the revised version of the Dutch edition:
De Westerweelgroep en de Palestinapioniers.
Nonconformistisch verzet in de Tweede Wereldoorlog
© 2015 Hans Schippers & Uitgeverij Verloren
www.verloren.nl

The publication of this book was made possible by:
Stichting Democratie en Media, Amsterdam
The Municipality of Rotterdam,
Gravin van Bylandt Stichting, The Hague.

Translator: R.J. Salfrais, Amsterdam
Editor: Jeanne Bovenberg-Meyers

ISBN 978-3-11-073682-3
e-ISBN (PDF) 978-3-11-058270-3
e-ISBN (EPUB) 978-3-11-058014-3

Library of Congress Control Number: 2018963877

Bibliographic information published by the Deutsche Nationalbibliothek
The Deutsche Nationalbibliothek lists this publication in the Deutsche Nationalbibliografie;
detailed bibliographic data are available on the Internet at http://dnb.dnb.de.

© 2020 Walter de Gruyter GmbH, Berlin/Boston
This volume is text- and page-identical with the hardback published in 2019.
Cover image: Group of Palestine pioneers in Barcelona in September 1944 (private collection). Printing and binding: CPI books GmbH, Leck

www.degruyter.com

Contents

List of Images —— VII

Timeline —— IX

Introduction —— 1

1   Palestine pioneers, an International Movement —— 6

2   Zionism and Palestine Pioneers in the Netherlands —— 15

3   Joop Westerweel and the Left-Wing Radical Milieu in the 1920s and 1930s —— 33

4   The Birth of the Westerweel Group —— 60

5   Expansion Activities and Reinforcements —— 85

6   A Second Hiding Operation —— 102

7   The Search for Safety —— 119

8   Hiding in Limburg, Germany and France —— 134

9   Westerbork and Beyond —— 159

10 The German Counter-Offensive —— 186

11 *Les Hollandais*, the Westerweel Group in France —— 216

12 After the Liberation —— 238

13 Helpers and Non-conformists —— 248

Bibliography —— 265

Glossary of Terms —— 270

Index —— 272

List of Images

Abbreviations

GFHA   Ghetto Fighters House Archives, Lochamei Hageta'ot, Israel
JHM   Joods Historisch Museum, Amsterdam, the Netherlands
MS   Mémorial de la Shoah, Paris, France
Priv. Coll.   Private Collection

Fig. 1   Palestine pioneers of the hachshara kibbutz of Lutsk in former Poland (now the Ukraine) in October 1934, with the first cow that they earned themselves (GFHA, no. 3629). —— 11
Fig. 2   A group of Palestine pioneers being trained in Den Bosch shortly after the First World War (JHM, no. F301856). —— 17
Fig. 3   The roll call of the members of the Werkdorp Wieringermeer when the Germans removed most of them on 21 March 1941 (GFHA, no. 41284). —— 24
Fig. 4   Mirjam Waterman in 1943 (Priv. Coll.). —— 31
Fig. 5   Wil and Joop Westerweel with two of their children in the 1930s (GFHA, no. 13749). —— 44
Fig. 6   'Workers' and 'helpers' of the Werkplaats in Bilthoven (ca. 1935). Kees Boeke at the top, in the middle, with Joop Westerweel to his left, in a dark suit. Philip Rümke is sitting three rows below Kees with his left arm on his knee. To the right of the flag we see Betty Boeke. Her daughter Candia, with a white collar is sitting in a row below her (GFHA, no. 33654). —— 52
Fig. 7   Bouke Koning at the beginning of the 1940s (GFHA, no. 13733). —— 66
Fig. 8   Jan Smit (GFHA, no. 62547). —— 70
Fig. 9   The Loosdrecht pioneers in spring 1942. Paul Sonnenberg fourth row far left, the Nussbaum sisters third row centre right. Manfred Paul with chequered shirt, third row left. Arthur Heinrich with chequered shirt, first row centre. His brother Josef with spectacles, rear row centre. Metta Lande, second row, far left. The Turteltaub brothers, second row centre, right and next to them Manfred Rübner with spectacles (GFHA, no. 13691). —— 73
Fig. 10   Kurt Hannemann (JHM, no. F000358). —— 88
Fig. 11   Antje Roos (left) with an unknown girlfriend in the Beth Chaloets (1942) (GFHA, no. 13600). —— 95
Fig. 12   Frans and Henny Gerritsen with their daughter Mieke. The photograph was made some years after the war (Priv. Coll.). —— 98
Fig. 13   The hachshara pioneers in Elden in the summer of 1942. Leader Werner Ahlfeld is standing on the far right (GFHA, no. 13606). —— 103
Fig. 14   A small group of the Gouda pioneers on one of the flat bottom boats (JHM, no. F008010). —— 107
Fig. 15   Eight pioneers of the Deventer Vereniging in Assen in February 1942. Several pioneers worked there temporarily in a paper factory. Max Windmüller can be seen on the far left, Kurt Hanneman is the fourth from the left, and Schraga Engel is far right (GFHA, no. 28578). —— 112
Fig. 16   Shushu Simon (JHM, no. F000357). —— 121

https://doi.org/10.1515/9783110582703-001

Fig. 17   Willy Gerler shown here during his hachshara training in Loosdrecht. He was one of the first pioneers who reached Spain at the beginning of 1943 (GFHA, no. 59387). —— 128

Fig. 18   Kurt Reilinger took over the leadership of the escape network in France from Shushu Simon in August 1943 (GFHA, no. 41266). —— 130

Fig. 19   Chiel Salomé took pioneers and other Jews who wanted to hide to Limburg and Frisia (GFHA, no. 13728). —— 136

Fig. 20   Max (Cor) Windmüller (right), here with fellow pioneer Harald Simon, accompanied pioneers from Brussels through France. Max died shortly before the liberation on one of the death marches from concentration camps. Harald Simon, who was arrested in Rotterdam in 1943, was murdered in Auschwitz (GFHA, no. 13672). —— 149

Fig. 21   The hachshara group, guarded by several MP's at work in the field outside the Westerbork camp (GFHA, no. 10169). —— 162

Fig. 22   Kurt Walter, shown here on his locomotive, was the main organizer of the escapes from Westerbork (GFHA, no. 33656). —— 174

Fig. 23   Lore Durlacher cared for pioneers in hiding and was involved in many escapes from Westerbork. The photograph is from 1942 (Priv. Coll.). —— 180

Fig. 24   Letty Rudelsheim, sitting far right, was the manager of the safe house in Rotterdam. The photograph was taken during a meeting of pioneers in Deventer at the beginning of the war (GFHA, no. 13608). —— 191

Fig. 25   Menachem Pinkhof around 1940 (Priv. Coll.). —— 207

Fig. 26   Members of the Armée Juive, with Avraham Polonski in the centre, in a post-war meeting with the Israeli Prime Minister David Ben-Gurion, to his right (GFHA, no. 13090). —— 219

Fig. 27   A maquis unit of the Armée Juive in the department of Tarn in 1944 (MS, no. MLX A11). —— 224

Fig. 28   Willy Hirsch, the organizer of the escape routes in Belgium and France, died in Bergen-Belsen, in 1945 (GFHA, no. 13716). —— 231

Fig. 29   In 1964, Wil Westerweel received the Yad Vashem award as 'Righteous among the nations' from Gideon Hausner, the prosecutor in the Eichmann trial and chairman of the board of Yad Vashem (GFHA, no. 41271). —— 246

Fig. 30   A group of Palestine pioneers in Barcelona in 1944. The group left Barcelona to go to Cadiz, from where they went to Palestine on the ship Guine. The little boy in the front row is Uri Durlacher (Priv. Coll.). —— 256

Map 1   The Westerweel group: hachshara training, important hiding centres, escape routes. © Peter Palm, Berlin. —— 85

Map 2   The Palestine pioneers in France: escape routes, maquis, OT worksites. © Peter Palm, Berlin. —— 216

Timeline

1899	Joop Westerweel born in Zutphen.
1910	First Palestine Pioneers in the Netherlands start agricultural training.
1923	Westerweel gets job as a teacher in the Dutch Indies. He is imprisoned for several months when he refuses to do his militia service. December 1924 he is expelled to the Netherlands.
1925	He works as a teacher in Amsterdam. In 1932 Westerweel and his wife Wil transfer to the Kees Boekeschool in Bilthoven. Active in various left-wing movements.
1933	Adolf Hitler takes power in Germany.
1934	Opening of the Werkdorp Wieringermeer for young German-Jewish refugees.
1939	Training for pioneers starts in Loosdrecht Pavilion. In total over 800 mainly German and Austrian pioneers stay in the Werkdorp and other similar facilities.
May 1940	German invasion of the Netherlands. Joop Westerweel headmaster at the Montessori School in Rotterdam.
Nov 1940	Jewish civil servants get fired. Anti-Jewish agitation by Dutch Nazi's.
1941	Februari staking (February Strike) in and around Amsterdam against persecution of Jews. Formation of the Jewish Council.
1941	Process of isolation and discrimination of Dutch Jews continues. Jews are forbidden to visit parks, cinemas and use public transport.
1942 May 3rd	Jews are forced to wear a yellow Star of David.
1942 July 15th	First Jews are transported from Amsterdam via Westerbork to Auschwitz.
1942 End of July/ early August	Westerweel Group was founded in Loosdrecht with Mirjam Waterman, Menachem Pinkhof, 'Shushu' Simon, Joop Westerweel, Bouke Koning, and Jan Smit as core members.
1942 Middle of August	Start of hiding operation of 50 pioneers from Loosdrecht.
1942 October	Attempt to bring eight pioneers to Switzerland fails due to betrayal.
Late 1942/early 1943	Increasing number of pioneers calls for help from the Westerweel Group to go into hiding.
1943 Spring	Hiding places in the North of Limburg become available. Document forger Frans Gerritsen joins the group. Start of the route to France.
1943 Summer/Autumn	Around 150 Palestine Pioneers travel as workers of the Organization Todt to France. Start of collaboration with the Armée Juive to escape to Spain.

X — Timeline

1943 Autumn	Betrayal in Rotterdam. Ten people get arrested among whom Wil Westerweel, Joop's wife.
Early 1944	First successful crossing of a larger group of pioneers to Spain. In total some 70 pioneers will reach Spain.
1944 March	Joop Westerweel and Bouke Koning are arrested at the Belgian border in Budel. 22nd of August 1944 Joop Westerweel is executed in Vught.
1944 May	Due to betrayal over twenty members of the Westerweel group are arrested in separate cases in Paris and the Netherlands.
1944 September	Most of the southern part of the country is liberated. The Westerweel Group ends its activities. Local resistance groups take over most of the care of some 100 pioneers.
May 5th 1945	Capitulation of the German forces in The Netherlands.

Introduction

During my master's thesis research in 1973 into the history of the left-wing socialist *Mapam* party, I happened to meet several former members of the Westerweel group in kibbutz Yakum. I was familiar with the name Westerweel group from the publications of Prof. J. Presser and Prof. L. de Jong as one of the few resistance organizations in the Netherlands in which Jews and non-Jews worked together during the Second World War.[1] However, the stories told to me by several former members of the group—about the hiding operations, escapes from Westerbork and journeys over the Pyrenees to Spain—were new to me and extremely fascinating.

Their stories, moreover, were not triumphant, but rather subdued and permeated with the realization that many comrades had fallen into German hands. The non-Jewish helpers were always mentioned with respect. Without them, going into hiding and carrying out the activities that followed later would not have been possible.

In Yakum, I was also encouraged, as a prospective historian, to consider researching this subject after my graduation. I agreed that this would indeed be interesting, but as we know, practical matters often stand in the way of the realization of such resolutions. Forty years later, here is the book.

The main reason for describing the history of the Westerweel group is its unique character. The group's members and helpers, Jewish and non-Jewish had different social backgrounds and political orientations. A second reason is that there was no fully researched monograph about the Westerweel group's activities when I started work on this book. I wanted to correct this omission.

Practical Problems

The Westerweel group was formed in the summer of 1942 when several leaders of the Palestine pioneers in the *Jeugdalijahhuis* (Youth Aliyah house) in Loosdrecht decided not to report for transport to Westerbork, but to go into hiding, together

---

[1] Prof. Dr. L. de Jong, director of the Netherlands Institute for War, Holocaust and Genocide Studies in Amsterdam, *Geschiedenis van het Koninkrijk der Nederlanden in de Tweede Wereldoorlog* (From now on: De Jong, *Koninkrijk*), (Den Haag: Martinus Nijhoff, 1969–1991). Dr. J. Presser, *Ondergang* (Den Haag: Staatsuitgeverij, 1965), of the about 140.000 Jews in the Netherlands in 1940 more than 100.000 (between 72% and 75%) were killed during the war. De Jong, *Koninkrijk*, dl. 8 tweede helft, 843–845.

with the residents of the house. They approached several non-Jews to find sufficient hiding places. One of them was Joop Westerweel, who quickly became the main organizer and inspiration of the operation. Later, others—both non-Jews and Jews—joined the Westerweel group.

When mentioning this name, the historian is confronted with a problem. In the middle of the 1990s, a documentary about the Westerweel group that was transmitted on Dutch television caused a great deal of commotion in the group. Some people thought that the documentary downplayed the part played by non-Jews in the resistance activities of the group, and assigned too great a role in these activities to the Jewish Palestine pioneers. This question, which mainly stayed behind closed doors and was also resolved there, was directly related to the name and composition of the group.

The name *Westerweel group* was first used in the 1960s, after the formal opening of the 'Joop Westerweel wood' in Israel. This name did not exist before then or during the war. Several former members of the group, both Jews and non-Jews, initially had reservations about the new name. They accepted it, however, as a token of recognition for Joop Westerweel, 'the engine' of the group, who paid the ultimate price for his efforts in front of a German firing squad.

In reality, however, the resistance group consisted of two or actually three nuclei, the group of non-Jewish helpers with Joop Westerweel as informal leader, the *chalutzim* (or Palestine pioneers) organized in the *Hechalutz* and the 'Haarlem' group under Frans Gerritsen. The different sections frequently cooperated. However, especially the Palestine pioneers, with their own organizational structure, and the Gerritsen section also had their own activities, separate from the Westerweel group.

The time factor also played an important role in assessing the part played by each section. For example, in the initial phase in August 1942, the group around Joop Westerweel played the main role in terms of hiding people and tending to the needs of the young pioneers from Loosdrecht.

From the end of 1942, the part played by the Palestine pioneers in the activities increased. Some months later, Frans Gerritsen and several assistants also started to work in the group. The group's work was now mainly aimed at getting as many pioneers in hiding as possible to France, and from there to Spain. Some of these pioneers had fled from Westerbork with the help of the Westerweel group.

To put it briefly, the part played by the group around Joop Westerweel was much greater in the autumn of 1942 than it would be a year later, when Frans Gerritsen and the Palestine pioneers had taken over several tasks in the growing number of activities. Partly as a result of the arrests of several members of the Westerweel group in the autumn of 1943 and spring of 1944, the part played by the pioneers and by Frans Gerritsen became more prominent, until the group ceased its primary activities in the autumn of 1944.

The Palestine pioneers were part of the Zionist movement, which had arisen at the end of the nineteenth century. This organization aimed to train young Jews for a life in agriculture in Palestine. The focus of the organization lay in Eastern Europe, where anti-Semitic sentiments were widespread. The Palestine pioneers had their own clearly socialist and non-conformist views.

In the Netherlands, the pioneer movement had become modestly active around the First World War. With the emergence of National Socialism in Germany in the 1930s, the number of Palestine pioneers in the Netherlands increased strongly, with the arrival of hundreds of refugees from Germany and Eastern and Central Europe. A limited group of young Dutch Jews also joined the movement.

The introductory chapter of this book discusses the ideological and social backgrounds of the Palestine pioneers, their role in the Zionist movement and their contribution to the resistance to the anti-Semitic measures of the Nazis.

Research Questions

This research attempt to find answers to several questions about the help offered by non-Jews to Jews and about the resistance of Jews to anti-Semitism.

One important question is what motivated the members of the Westerweel group to offer help to their Jewish fellow citizens? The answer to this question is placed in a larger framework in the concluding chapter.

A second issue has to do with the specific social character of the Westerweel group that manifests itself, for instancein the position of women and the pacifist attitude of some of its members.

Sources

Extensive use was made of the statements provided in the years 1954–1957 by members of the group regarding their wartime activities. These individual statements are lengthy responses—three to four pages, with a maximum of six pages—to standardized questions about this period. Because of the short interval between the war and the statements they seem highly reliable.

The answers (most of them in German, some in Dutch or Ivrit) are generally short and business-like, providing a good factual representation of the events in which the group's members were involved during the war period.

Other important sources include the interviews held by Kurt/Yigal Benjamin at the end of the 1980s with mainly non-Jewish members the Westerweel group.

Equally valuable were several interviews dating from the 1990s by Mirjam Pinkhof-Waterman and the interviews by Sytske de Jong for her M.A. thesis (2001) about the Westerweel group.

Furthermore, several biographies based on diaries, by for example Paul Siegel, Letty Rudelsheim and Hans Flörsheim, were of great importance.

Acknowledgements

Many people and institutions have helped and encouraged me in writing this book. I would hereby like to express my gratitude to them all.

First, I would like to thank Prof. Dr. Bart Westerweel, the son of Joop and Wil Westerweel, who received me hospitably at his home several times, made material available and gave critical comments on my texts.

Drs. Sytske de Jong gave me the material that she had collected for her thesis about the Westerweel group. This also included several important interviews with former members. Philip Rümke (a member of the group), Mieke Bouman-Gerritsen (daughter of Frans and Henny Gerritsen) and Guido Smit (son of Jan and Helga Smit) provided material and photographs and acted as second readers and commentators on the draft texts.

I also received comments from Dr. Bart van der Boom (University of Leiden) and Dr. David Wertheim (Menasseh Ben Israel Institute), Prof. Dr. Em. Hermann von der Dunk and drs. Annemieke Gringold (Jewish Historical Museum, Amsterdam).

Drs. Sierk Plantinga (National Archive) and Dr. Lieven Saerens (CegeSoma in Brussels) gave their reactions to several chapters. Dr. Daniela Hooghiemstra and Dr. Herman van Rens made several chapters available from their then as-yet-unpublished dissertations.

My special thanks go to Prof. Dr. Johannes Houwink ten Cate of the Netherlands Institute for War, Holocaust and Genocide Studies (NIOD), who provided the whole manuscript with detailed comments and remarks and referred me to other experts.

Prof. Dr. Marjan Schwegman, director of the NIOD, gave advice with regard to the approach and named possible sponsors. She also made me aware of the possibility of working on the book as an associate researcher with the NIOD. Dr. Eveline Buchheim assisted with the subsidy requests.

Itamar Lehmann translated several texts from Ivrit, and drs. Gerard Telkamp gave linguistic advice. At the initiative of drs. Katinka Zeven Chapters 5 and 6 were revised by the students of the MA translation in Theory and Practice at Leiden University.

I received financial support for the research from the Stichting Democratie en Media, the Municipality of Rotterdam and from the Gravin van Bylandt Stichting. I am thankful that De Gruyter Oldenbourg wanted to publish an English translation of my book in the Series New Perspectives on Modern Jewish History. There was a pleasant cooperation with Acquisitions Editor Dr. Julia Brauch and with Series Editor Prof. Dr. Cornelia Wilhelm.

I would also like to express my thanks to my former colleagues of the TIS group at the Industrial Engineering and Innovation Sciences faculty of the Technical University of Eindhoven. They allowed me to make continued use of a room and of the services of the library. Peter Smits ensured that the 'exotic' books (for the technical world, that is) that I requested actually also landed on my desk.

My wife Frank accompanied me several times on visits to (foreign) archives. She was also the level-headed, careful second reader and text advisor who protected me from mistakes. Those that remain are entirely my own responsibility.

1 Palestine pioneers, an International Movement

At the end of October 1944, the Portuguese ship *Guine* left the harbor of Cadiz in Spain, setting its course for Haifa in Palestine (which was under British mandate) with more than 430 Jewish refugees, including 55 so-called Palestine pioneers (in Ivrit, *chalutzim*). One of these pioneers was Paul Siegel. Helped by members of the Westerweel group, he had reached Spain after fleeing from Westerbork internment camp and enduring a grueling journey over the Pyrenees. Ten days later, he heard the lookout calling: 'Palestine ahead! Palestine ahead! We all rushed to the deck to miss nothing of this historic moment and to gaze upon the coast of Palestine, the country that had been so difficult to reach since the time of Moses. And, indeed, we saw the Carmel mountain range on the horizon'.

The arrival of a ship with refugees from occupied Europe in wartime was exceptional. There were therefore many people present to welcome the *Guine* and her passengers. *Jewish Agency* representatives asked Siegel and other pioneers to join in singing the *Hatikva*, the national anthem, there on the quay.

'There I stood on the deck with my throat choked with emotion and completely unable to sing. At last, we were in the country that, a year ago, I had not dared to believe I would ever reach. At the same time, I thought of all my comrades who would never reach Palestine. I thought of my loved ones, my parents and sisters, of whom I did not know if they were still alive.' Siegel's fellow passengers had similar thoughts: 'the Hatikva could hardly be heard on the deck'.[1]

The *Hechalutz*, the umbrella organization of Palestine pioneers, was a vital but in Western Europe lesser-known part of the Zionist movement, which was founded in 1897 by the Austrian-Jewish publicist Theodor Herzl (1860–1904). Because of increasing anti-Semitism, which destroyed all hopes of Jewish emancipation in Europe, Zionism set its sights on realizing a Jewish national home in the Turkish province of Palestine.

Zionism must be seen in the light of the overstrained nationalism that existed around 1900 in Europe. Herzl's organization can be seen not only as a reaction to this nationalism, but also as a part of it. Especially in Eastern and Central Europe, there were conflicts between all sorts of population groups that sought independence or the greatest possible autonomy. These nationalistic tensions contributed significantly to the outbreak of the First World War in 1914.

---

1 Paul Siegel, *Locomotieven trekken wagons 1933–1945* (Westervoort: Van Gruting, 2000), 184–185.

Some of these tensions were exacerbated at the end of the nineteenth century by the emergence of pseudo-scientific theories about the racial superiority or inferiority of population groups. Here, Jews were considered inferior, together with Slavic and colored peoples. In Germany, a 'nationalistic' variant—the German population as a racially homogeneous Germanic community—found quite some support. According to these views, Jews were not part of this nationalistic community.

At the end of nineteenth century, the large majority of European Jews lived in the multi-ethnic state of Russia, which also included part of Poland until 1919. In this authoritarian-governed country, there were large social and political tensions. The policy of Tsars Alexander III and Nicolas II was aimed at making Russia a Slavic unitary state, in which there was no place for minority groups such as Swedes, Germans and Jews.

With regard to the Jews (who, apart from a limited number of wealthy individuals, were allowed to live only in the so-called Pale of Settlement at the western part of the Russian empire), this led to a policy of forced assimilation, emigration and impoverishment through economic marginalization. The secret police supported this policy by organizing pogroms. These anti-Jewish riots made thousands of victims, mainly in the period of 1903–1906—after the Russian defeat in the war against Japan and the failure of the 1905 revolution.

The general anti-Semitic climate in Eastern and Central Europe led to the emigration of many Jews. The large majority of emigrants went to the United States; the rest went to countries such as Argentina and South Africa. Between 1899 and 1914, 1.3 million Jews emigrated from Russia. Most of them left between 1903 and 1906, when 400,000 Russian Jews settled in America. Under the influence of Zionism, the Jewish population in the Turkish province of Palestine also increased. From a first more or less reliable census in 1844, it appeared that about 17,000 Jews lived in Palestine. This number was 24,000 in 1882 and grew to approximately 85,000 in 1914.[2]

Socialist Zionism and the Inverted Pyramid

The new Jewish inhabitants often used Arab workers for their agricultural projects. This practice was rejected by the socialist Zionists, who thought that this

---

[2] Walter Laqueur, *A History of Zionism* (London: Weidenfeld & Nicolson, 1976), 60, 75–80. See also Zvi Gitelman, *A Century of Ambivalence* (New York: Yivo Institute for Jewish Research, 1988), 2–19.

work also had to be done by Jews. Zionism began as a liberal movement. There were no socialists at the first congress in Basel, but socialist Zionists played an important part in subsequent meetings.

Leaders of the movement, such as Nahman Syrkin and Ber Borochov, were significantly influenced by the Marxist views of the left-wing Russian organizations. They doubted however, whether this ideology could offer a solution for the predicament in which the Jewish proletariat in Eastern Europe lived. Prominent Marxists recognized the problem, but they insisted that the coming revolution was the answer. The anti-Semitism of the mobs manipulated by the reactionary forces would then disappear.

In 1905, the socialist Zionists united in the *Poale Zion* organization. A core part of their ideology was to realize a change in the professional structure of the Jews from Eastern Europe who settled in Palestine. These Zionists compared this structure with an inverted pyramid featuring a broad layer of (small) independent entrepreneurs at the top and a limited group of farmers and industrial workers at the bottom. This undesirable situation had arisen because European Jews had, for centuries, been forbidden to own land, to become member of a guild or to exercise certain professions. *Poale Zion* maintained that this had to change in Palestine. Here, Jewish immigrants would 'have to conquer work'. This meant that they had to form the collectively organized farming population and the industrial proletariat.[3]

To prepare themselves for these tasks, various Jewish organizations—both socialists and others—proposed a training program prior to emigration to Palestine. This preparation period was called *hachshara*, in which the youth movement would have to play an important part.

The Jewish Youth Movement and the 'New Style'

In the beginning of the twentieth century, the youth movement in Europe had fallen under the influence of a renewal that originated in Germany, at the initiative of the young people themselves. Led by peers instead of older people, the *Wandervogel (Wanderers)*, as they called themselves, made journeys in the countryside on weekends and on holidays. This new youth movement developed its own culture, with meals around the campfire, folk dancing and abstinence from alcohol and tobacco as its characteristics.

---

[3] Laqueur, *A History*, 272–278. See also Gitelman, *A Century*, 26–28.

More important than these material matters was the creation of a 'new style', as the American-German historian Walter Laqueur called it. Activities such as striving for independence from bourgeois society, self-government and tackling challenges were prominent objectives. A more equal place for women and open discussions about sexuality were also new aspects.[4]

The example of the *Wandervogel* was also adopted by the Jewish youth movement in Germany, and later elsewhere. Moreover, the creation of separate Jewish organizations was stimulated by the fact that Jews were often not welcome in the *Wandervogel*, just as in other associations and clubs. This selectivity in admission gave an impulse to Zionism, which found its strength in forming its own organizations. In 1913, a first Jewish youth association arose, called *Blau-Weiss* (Blue-White), the colors of the Zionist flag.[5] Other Jewish organizations followed later, with a wide range of liberal, socialist, Zionist, non-Zionist and religious ideologies, just as in the general German youth movement. In this respect, the socialist Zionist *Habonim* movement, which arose from a merger in 1933, is especially important.

After losing the First World War, Germany experienced strong growth of right-wing and extreme right-wing organizations, often featuring a clearly anti-Semitic agenda. This gave the Jewish movements more the character of a home and a safe harbor for their members. When the situation for Jews in Germany worsened in the 1930s, the organizations concentrated more on the *Hechalutz*, the training of pioneers for Palestine, and the associated *hachshara*, the preparation for work in agriculture and industry in that country. For example, *Habonim* set up camps offering training not only in agriculture, but also in the use of Ivrit (modern Hebrew) and knowledge of Palestine.[6]

The 'new style' of the modern youth movement did not remain limited to Germany. The strongest influence in Jewish circles was probably seen in *Hashomer Hatzair* (Young guards). This movement was founded in Galicia, a region spreading over parts of Ukraine, Poland and Czechoslovakia, shortly before the beginning of the First World War. Several members who had fled to Vienna during this war became acquainted with the ideas of the *Wandervogel*, which they then introduced into their own circles.

Hashomer Hatzair distanced itself clearly from the Jewish establishment and did not tie itself to any one political party or organization, but operated autonomously, driven by its own framework. Ideologically, elitist romantic Nietz-

---

4 Walter Laqueur, *Young Germany* (New Brunswick/London: Transaction Books, 1984), 3–31.
5 Laqueur, Young Germany, 76.
6 Jutta Hetkamp, "*Die jüdische Jugendbewegung in Deutschland von 1913–1933*" (PhD diss., University Essen, 1994), 42–68, 95–155.

schean views dominated at the beginning. Around 1920, the movement accepted Borochov's socialist Zionist model as its guiding principle. However, the elite character remained intact. *Hashomer Hatzair* saw itself as the vanguard of the Jewish proletariat that had the double task of realizing a changed social structure of the Jewish community in Palestine and of contributing to the class struggle. At the beginning of the 1920s, the first members of the movement settled in Palestine to organize collective farms: *kibbutzim*.

At the end of the First World War, other youth movements arose in Eastern Europe. They differed in philosophy of life, and also according to the region where they were active. The main movements were the following:

Dror (Freedom), which arose in the Ukraine, had Kiev as its most important center. When Zionist activities were forbidden by the government several years after the Russian Revolution, the members moved to Galicia in Poland. From here, *Dror* developed into the largest Zionist socialist youth movement in Poland, with a membership that clearly remained behind that of the non-Zionist youth chapter of the *Bund (Jewish workers)*, which tried to achieve the greatest possible autonomy of the Jews in Poland.

Gordonia, which was founded in 1923 and was also active in Romania, had a comparable political direction as *Dror*. The movement was named after A.D. Gordon, a pioneer of the 'conquering work' ideology. The more populist *Gordonia* movement rejected the class struggle and emphasized the 'idealism of the productive people'.

Betar, an acronym of Brit Josef Trumpeldor, was founded in Riga in 1923.[7] The organization, which represented the other end of the political spectrum, developed during the course of the 1920s into the youth movement of Vladimir Jabotinsky's right-wing Union of Revisionist Zionists. *Betar* had many members in the Baltic States and Eastern Poland. During the 1920s, umbrella religious youth organizations were also founded in Poland. The main organizations were *Mizrachi* and the strictly orthodox *Agudah*.[8]

Because many Jews left Poland to go to Western Europe during the 1920s and 1930s, all of the above-mentioned organizations had chapters in these countries as well.

---

[7] Joseph Trumpeldor (1880–1920): After his courageous behaviour during the Russo-Japanese war (1904–1905), he was the first Jewish officer in the Russian army. He later went to Palestine and died while defending a kibbutz in north Palestine against Lebanese infiltrators. He played a role in setting up the *Hechalutz* and was regarded as an example by both left-wing and right-wing youth movements.

[8] Ido Bassok et al., "The Youth Movements, Dror, Gordonia, Betar, Agudah Yisrael, Mizrahi articles," https://www.yivoencyclopedia.org., accessed 6 januari 2014.

These youth movements were all influenced in some degree by the 'new style' of the German youth movement. They organized their activities themselves, for example, and maintained an autonomous position with regard to the political organizations to which they were linked. This also applied to *Betar*, which kept its distance from the Revisionist Party. Although it can be said that the religious organizations operated with a certain autonomy, they did not adopt any substantive aspects of the modern youth movement.

During the 1920s and 1930s, the Jewish youth movement developed into a driving force behind Zionism, often expressing sharp criticism of the Zionist movement's leaders, who were accused of sluggishness and a middle-class mentality. The youth movement asserted that the Palestine pioneers were the people who were serious about the ideals of Zionism. They were the ones who went on *Aliyah* —literally: going up, in the meaning of emigration to Palestine —in contrast to many (board) members of the Zionist organizations.

Fig 1: Palestine pioneers of the hachshara kibbutz of Lutsk in former Poland (now the Ukraine) in October 1934, with the first cow that they earned themselves

Palestine Pioneers and *Hachshara*

Pioneer training courses for Palestine were set up in Eastern Europe from about 1905. They were mostly initiatives of local Zionist organizations. An umbrella organization, the *Hechalutz*, arose in Russia after the October revolution. An important impulse for emigration was given by the Balfour Declaration of November 1917, in which the British government promised the establishment of a Jewish Home in Palestine.

Besides knowledge of the country and Ivrit, the preparatory *hachshara* was part of the training. The *hachshara* had two forms: the individual (whereby pioneers learned practical knowledge alone or in small groups with farmers or handicraft companies) and the collective (whereby dozens of pioneers worked in their own kibbutz or large farm).

As a result of the spontaneous initial period, the first *Hechalutz* had no clear ideological character. However, the organization quickly took on a more socialist orientation. The movement's focus lay in the Soviet Union. In the mid-1920s, however, the communist party tightened the ideological reins, making it clear that Zionism should be regarded as a nationalistic deviation. The network of *hachshara* institutions was dismantled or absorbed into the communist party's agrarian training program of the *Jevsektia* (Jewish department). Several leaders of the *Hechalutz* disappeared to Siberia.[9]

Some of the pioneers then left for Poland, which had become independent in 1919, and the movement gained a new focus. The new state had a Jewish minority of approximately 10%, which lived mainly in Eastern Poland, the core of the former Pale of Settlement, and in the big cities. The mostly Yiddish-speaking Jews formed a clearly distinguishable population group, who were confronted with serious discrimination during the first years of the state. They and other minority groups were regarded as non-national elements. Moreover, the Jews were assumed to sympathize with the communist Soviet Union. During this period, there was great pressure to emigrate. The pioneer work, which prepared people for emigration, offered an alternative to this distressing situation. The young people who joined felt protected in their own circle. The possibility of a Jewish state promised in the Balfour Declaration functioned as a trigger.

Poland remained the most important base for the *Hechalutz*. However, depending on the political and economic climate in the country, the number of people that enrolled for the pioneers' work varied greatly. The pioneer movement grew especially strongly in the 1930s, with the emergence of nationalistic parties and increasing discrimination. In 1935, the total membership of the *Hechalutz* was around 30,000—of which, 8,500 lived in *hachshara kibbutzim*. In this spectacular growth, pragmatism also played a part. Many Jewish youths wanted to leave Poland at any price. The *Hechalutz* offered the possibility of obtaining a

---

9 Gitelman, *A Century*, 108–121.

certificate (entry visa) for Palestine. Once they arrived there, some of them broke their ties with the kibbutz movement and went to the cities or left the country.[10]

Important organizations of Palestine pioneers with a preparatory practical training were also set up in the 1930s in Lithuania, Romania, Czechoslovakia and Germany, which was mentioned above. Generally speaking, the activities in the above-mentioned countries were, however, more limited than in Poland. Outside Europe, there were *Hechalutz* organizations in Iraq, South Africa, Argentina and the United States, among others. At its highest point in 1933, a total of more than 80,000 pioneers were registered at the movement's main office in Warsaw. Many applicants were not able to follow a *hachshara* training course because there were long waiting lists.

The Palestine Pioneers and Palestine/Israel

The left-wing Jewish youth movements played an important part in Palestine and during the first decades of Israel, which became an independent state in 1948. The number of members of *kibbutzim* rose from 2,000 in 1927 to 24,000 in 1940—and then to 65,000 in 1950, which represented 7.5% of the population. They did not all come from the youth movement and they had also not all followed a *hachshara* training course, but most of them had.

The youth movement's members provided an important part of the framework for building the country. They were people for whom, as Paul Siegel in the quotation at the beginning of this chapter, only one thing counted during and after the severe circumstances of the Second World War: the *Aliyah* to Palestine and the establishment of the state of Israel. Pinhas Lavon, the leader of *Gordonia*, fulfilled such functions as Minister of Defense and chairman of the union movement *Histadrut*. The leader of *Dror*, Yitzhak Tabenkin, was a member of the Knesset (the Israeli parliament) and leader of the kibbutz movement. Meir Ya'ari of *Hashomer Hatzair* founded the left-wing socialist *Mapam party* in 1948 and was then a member of the Knesset until 1973.

During the War of Independence and the wars of the 1960s and 1970s, the *kibbutzim* youth formed the backbone of the Israeli army. The *kibbutzim* had an important economic position, but they also formed a testing ground for social experiments. The values of the left-wing youth movement—equality, sobriety and a strong work ethic—were, to an important degree, also the values of the state. The left-wing part of the kibbutz movement furthermore pressed for a territorial

---

[10] Gitelman, *ibid.*, 52–68.

compromise/agreement with the Palestinians. This influence only subsided gradually during the 1980s when Israel became a modern capitalistic state, with the ensuing substantial differences between rich and poor.

2 Zionism and Palestine Pioneers in the Netherlands

The circumstances of the Jewish community in the Netherlands were hardly comparable with those in Eastern and Central Europe. Jews received equal citizenship rights in the Netherlands in 1796 during the time of the French rule. Germany (1871), Austria-Hungary (1867) and Russia (1917) followed quite some time later. The granting and implementation of citizenship rights did not go easily in the Netherlands. There was widespread resentment against Jews, but this anti-Semitism was latent rather than publicly expressed.

During the nineteenth century and the beginning of the twentieth century, Dutch Jews became socially assimilated to an important degree. They were active as entrepreneurs and in free professions such as doctors, lawyers and journalists. Jews also played an important part in socialist and liberal parties. The sizable proletariat in the big cities was, however, still mainly concentrated in traditionally 'Jewish' professions such as the itinerant trade, the clothing industry and the diamond industry. From the 1920s, there was increasing anti-Jewish agitation arising from growing international right-wing radicalism. However, the number of members of various organizations with this ideology, such as the Nationaal-Socialistische Beweging (NSB, Dutch Nazi party) and the fascist Zwart Front (Black Front), remained relatively low.

The Zionisten Bond (Union of Dutch Zionists)

Partly as a result of these circumstances, the Dutch Zionisten Bond (NZB), which was founded in 1899, had only a limited membership. Of the roughly 130,000 Jews in the Netherlands in 1935, approximately 3,000 were members of the NZB. This increased to more than 4,200 in 1939. Members came largely from the well-to-do middle class; the organization had hardly any members among craftsmen and small merchants. Important people in the Union were the founder, the insurance expert Nehemia de Lieme, the banker Jacobus Kann, professor of ancient history and philosophy Dr. David Cohen and the lawyer Abel Herzberg. Rabbis Simon de Vries and Aaron Davids were also members. They played an important part in the religious Zionist *Mizrachi*, which was part of the NZB. Finally, in 1935, a chapter of the socialist *Poale Zion* with about 500 members was established, which, after the necessary bickering, also joined the NZB. The most influential man in their circle was the Marxist economist Sam de Wolff.[1]

Gouda and the Deventer Vereniging (Union)

Considering the limited size of the Dutch Zionist movement, it is remarkable that a training organization for Palestine pioneers was started in Gouda as early as in 1910. Although the initiator was Jacobus Kann, the training course floundered. Up to 1918, only three pioneers were trained at the Ida-Maria State estate, which was purchased for this purpose. This dismal result was also partly due to the poor accessibility of Palestine during the war.

The initiative was then taken over by Rudolf 'Ru' Cohen, who lived in Deventer and was the owner of a furniture shop and brother of David Cohen. Together with Schlomo and Emile Fischer of the Eastern-Jewish community living in Scheveningen (near The Hague), he founded the 'Vereniging tot Vakopleiding van Palestinapioniers in Nederland' (Association for the vocational training of Palestine pioneers in the Netherlands) in 1918. Under the name of the Deventer Vereniging (Deventer Union), the institution became a household name in Zionist Netherlands. Cohen's organization placed the pioneers individually with farmers to learn all sorts of agrarian skills. Somewhat later, they also worked from *hachshara* centers in Marum in Groningen, in Twello and Winterswijk in Gelderland and in Den Bosch in Brabant, among many others.[2]

Almost all of the pioneers came from Eastern Europe, and most had a very different mentality and background compared to Dutch Zionists. The eastern European Jewish pioneers came from the milieu of workers and small traders, and often arrived destitute in the Netherlands. In the countries from which they came, open anti-Semitism was a bitter reality. For them, Palestine was the Promised Land where they could build a new life.

The pioneers therefore worked very hard to become adept at the heavy farm work, with which most of them had no previous experience. They worked long days, especially in the summer. The reward for this hard work was meagre. During the first months, the pioneers did not even receive pocket money and, after-

---

1 Ludy Giebels, *De Zionistische beweging in Nederland 1899–1941* (Assen: Van Gorcum, 1975), 165–199.
2 Giebels, *De Zionistische*, 124, 125, 193, 194.

Fig. 2: A group of Palestine pioneers being trained in Den Bosch shortly after the First World War

wards, skilled workers got no more than half the salary of a farm worker, who also did not keep much over of his wages.³

To be able to cope with the isolation in which they found themselves in the Dutch countryside, the *Hechalutz* Netherlands movement organized monthly meetings. Here, the pioneers could let off some steam and talk about their problems and ambitions. Their efforts, eagerness to learn and energy ensured that they had no problem finding training places to learn about farming. The Zionist idealism impressed many people with whom they came in contact. As far as one can ascertain, the pioneers had a good name with the Dutch farmers.

There was, however, quite a lot of friction with the NZB board. They did not understand the more emotional eastern-Jewish mentality of the pioneers, who were mostly confirmed socialists and, moreover, were often brought up in the 'new style' of the Zionist youth movements. The idiosyncratic and seemingly arrogant idealism of the pioneers upset several board members. They had hardly any appreciation for socialism and the collectivistic ideal of the kibbutz. The *Joodse Wachter* (*Jewish Guard*), the NZB's newspaper, regularly published rather scornful articles about the pioneers who seemed to be very satisfied with them-

---

3 Yigal Benjamin, *They were our Friends* (Tel-Aviv: Association of former members of the Hachshara and Hechalutz Underground in Holland, 1990), 9–11.

selves. 'Nowhere else were the conditions so difficult, not even in Eretz Israel', wrote former Palestine pioneer Yigal Benjamin, perhaps with some exaggeration in his commemorative book about the *Hechalutz* in the Netherlands.[4]

The NZB board was also not very forthcoming financially. The treasurer maintained that the pioneers had to stay in the Netherlands longer, if necessary, to save for the travel costs to Palestine (about 100 guilders) from their pocket money. The NZB did not have this money, and the Bond did not want to set up a fund for it.

In a reaction, a pioneer pointed out that it was certainly not the intention 'to 'paint haloes around the heads of' the *chalutzim*'. However, to treat them as beggars, to whom you could toss some petty cash, was an insult. 'The work in Israel cannot be done without the chalutzim, and the chalutzim need the Zionists just as much'[5].

The foreign pioneers mainly did as they liked, and paid little attention to the NZB board—and the feeling was mutual. The Bond did not attempt to get more grip on the organization until more Dutch youths applied for a *hachshara* training course in Deventer during the 1930s. These young people came from the umbrella *Zionistische Joodse Jeugd Federatie* (JJF, Federation of Zionist Youth), which was founded in 1920. Partly because they were confronted with increasing anti-Semitism, they took a more active stance than the more cautious adult members.

Despite this difficult relationship, several hundreds of pioneers trained in the Netherlands did reach Palestine, mainly through the efforts of Ru Cohen's Deventer Union. In 1938, more than 200 people were present in kibbutz Ein Harod to honor Cohen and his wife and to receive from him the gift of a Friesian pedigree bull with the name of Ru, which Cohen had taken with him from the Netherlands for the *kibbutz*.[6]

In the 1930s, the religious Zionist *Mizrachi* movement founded two *hachshara kibbutzim* in Franeker and Beverwijk, respectively. For the strictly orthodox pioneers of the *Agudah*, a kibbutz was founded later in Enschede. All three farms worked on a collective basis for practical religious reasons. This collectivism was obviously independent of the socialist philosophy of life of most pioneers in the Deventer Union.[7]

---

4 Benjamin, *They were*, 11, 12.
5 Giebels, *De Zionistische*, 192–196.
6 Giebels, *ibid.*, 178, 179, 195–199.
7 Benjamin, *They were*, 13, 14.

The *Werkdorp Wieringermeer* and the *Paviljoen* in Loosdrecht

Adolf Hitler's rise to power at the end of January 1933 triggered the start of an influx of Jewish refugees to the Netherlands. In the first year alone, there were 4,000 refugees. Cohen and his union were soon no longer capable of organizing the shelter of *hachshara candidates*. In 1934, therefore, the *Stichting Joodse Arbeid* (Foundation for Jewish Labor) was founded; at the beginning of October of the same year the foundation opened a 300 hectare *Werkdorp* (work village of about 740 acres), in the recently drained polder of Wieringermeer. Two work camps that were used to house workers during the draining were made available as the base for this foundation. In the *Werkdorp Wieringermeer* 150 Jewish refugees from Germany and Poland and later Austria could be trained in farming work and handicraft.

Most of the refugees were initially non-Zionistic, but wanted to learn a trade to be able to work in the country in which they would eventually settle. Emigration was a requirement of the Dutch government, which was concerned that the refugees would end up on the labor market. Therefore, small groups of pioneers left the *Werkdorp* at set times to establish themselves in Palestine or elsewhere. In July 1939, the ship *Dora* left for Palestine with a total of about 500 mostly German-Jewish refugees. The passengers also included dozens of pioneers from the different training courses. However, because it became increasingly difficult at the end of 1930s to get certificates for Palestine or entry visas for other countries, there were still around 300 pioneers in the *Werkdorp* in May 1940.

There were about 520 more pioneers participating in six other *hachshara* courses in the Netherlands, making a total of more than 800, most of whom came from Germany, with the rest coming from Austria and Poland.

In June 1940, the management of all these training courses was transferred to the *Joodse Centrale voor Beroepsopleiding* (JCB, Jewish Center for Vocational Education). In February 1941, the JCB became part of the *Joodse Raad* (Jewish Council), which was set up by order of the Germans.[8]

From September 1939, one of the training courses was in the *Paviljoen Loosdrechtsche Rade*. The *Paviljoen* (Pavilion), as it was mostly called, had previously been a home for children with behavioral problems. The small building was then purchased by the Jewish family Deutsch, who had made it available to the *Jeugdalijah*. This organization was founded in 1933 by Recha Freier, the wife of a Berlin

---

[8] H.B.J. Stegeman and J.P. Vorsteveld, *Het Joodse Werkdorp in de Wieringermeer* (Zutphen: Walburg Pers, 1983), 34–80.

rabbi. She foresaw that a normal life for Jews in Germany would no longer be possible after the Nazis gained power.

The *Jeugdalijah's* goal was to get as many children and young people as possible out of Germany and later also out of Austria. The primary destination was Palestine. However, because of the entry restrictions imposed by the British in the *White Paper*, for example, children also went to other countries. Erika Blüth, who came from Germany, was made leader of the *Jeugdalijah* in the Netherlands, together with a Dutch assistant.

The children who arrived in the Netherlands after the *Kristallnacht* of 10 November 1938 (see also Political Activities in Chapter 3) were initially housed in an empty school building in Amsterdam. The inhabitants, who varied in age from 13 to about 18 years old, were then divided over the *hachshara* training courses. Twenty of them also went to the agricultural training center in Gouda, which was refurbished in 1937 as a result of a large donation.

In Loosdrecht, the leader was initially Jacov Zurawel, who came from Palestine, and was assisted by several assistants. Zurawel was sent to the Netherlands as *shaliach* (advisor) to lead the practical side of the *hachshara*. This started during the autumn of 1939, when the pioneers were housed individually with farmers in the surroundings to assist with all sorts of activities. Two young men followed a training course to become blacksmiths. Most girls learned how to milk cows and did other light farming work or helped in the household of the farm.

After the German invasion in May 1940, the group was briefly evacuated because of the threat of inundations. After the capitulation, the pioneers quickly returned. Director Zurawel, who had an English passport, had moved to England with his family in May 1940. The Dutchman Lodi Cohen, who was 23 years old, then became the leader. His assistants Channa de Leeuw and Betty Britz were mainly occupied with domestic matters in the *Paviljoen*. Cohen had followed a rabbinical study and regularly taught religious subjects in Loosdrecht.[9]

Later, Menachem Pinkhof and Joachim Simon were made youth leaders. Pinkhof was born in Amsterdam in 1920 and was a member of the *Mizrachi*, the religious Zionist youth movement. He left this movement shortly before the war and joined the socialist *Hechalutz* movement. He studied civil engineering at the then MTS (Polytechnic College) in Amsterdam, and had requested a Palestine certificate at the beginning of 1940. This was refused, however, because he still had to go into military service. In May 1940, Pinkhof had made plans to flee to France, together with several others, but this was not successful. After he gradu-

---

9 Mirjam Pinkhof and Ineke Brasz, eds., *De Jeugdalijah van het Paviljoen Loosdrechtsche Rade, 1939–1945* (Hilversum: Verloren, 1998), 15–59.

ated, he was unable to find a job, partly because it was now forbidden for Jews to work for non-Jews. In 1941, at the request of the *Jeugdalijah*, he went to Loosdrecht.

Joachim 'Shushu' Simon was born in Berlin in 1919 into a liberal-Jewish family. His mother died shortly after his birth. Together with his older brother Jakob, he was raised by his father, with the assistance of an aunt. The family were confirmed Zionists and Simon went to a private modern Jewish school, where Ivrit was taught. He was also a member of *Kadima*, a socialist Zionist youth movement. This would merge at the beginning of 1933 with another left-wing youth movement to become the *Habonim* movement.

When his father remarried in 1932, 13 year-old Simon went to Frankfurt to live with his aunt and grandmother. His brother went to a *Habonim* pioneers camp and from there to Palestine. In 1936, 'Shushu' had to leave secondary school because he was Jewish. He returned to Berlin to finish his education at a Jewish school. An attempt to go Palestine afterwards failed because he did not have a certificate. To improve his chances of getting this document, Simon enrolled in a *hachshara* program with 90 other *Habonim* members in the Elgut camp in Silesia in the autumn of 1937. Here, he was arrested in November 1938 after the *Kristallnacht* and locked up in the Buchenwald concentration camp. When he was released after more than a month, the German police had closed all pioneer camps.

Many pioneers decided to continue their *hachshara* programs abroad. Shushu Simon landed up in Deventer with his head shaved bald, and with no more possessions than the clothes he was wearing. He worked on a farm for some time, but a new attempt to get a certificate for Palestine failed. At the beginning of 1940, Simon taught Ivrit and was a group leader in the *Werkdorp Wieringermeer*. Shortly afterwards, he was given the same position in Loosdrecht. He was also a board member of *Hechalutz* Netherlands and contact of Erika Blüth of the *Jeugdalijah*, who later also had a position with the Jewish Council in Amsterdam.[10]

Shushu Simon had an extravert personality, recited poems, told jokes and led the folk dancing. Menachem Pinkhof was more the type of a level-headed *no-nonsense* go-getter. Despite these differences in character, they respected each other and worked well together.

---

[10] Ghetto Fighters House Archives, Lochamei Hageta'ot, Israel (From now on: GFHA), Cat. no. 173. Several undated statements (Verklaringen) by Menachem Pinkhof around 1960. The most extensive of these was an interview on 23 May 1960 with Haim Avni of the Institute of Contemporary Jewry of the Hebrew University in Jerusalem. Pinkhof wrote several texts in preparation and/or as supplements to this interview and for other occasions. (From now on: Verklaringen Pinkhof)

The Loosdrecht Pioneers in Wartime

In the *Paviljoen*, the group housed there after May 1940 had undergone several changes. In the summer after the capitulation, the *Paviljoen* took in more than 20 inhabitants from another pioneer center. About 70 pioneers remained in Loosdrecht—and that was too many. Several of them had to be housed in a guesthouse or with private individuals. A small group of mostly youthful pioneers left in the course of 1941 to other training courses; several joined their parents.

In the summer of 1942, more than 30 male and almost 20 female pioneers lived in Loosdrecht, including the leaders. A few of them were orthodox and kept to the dietary laws, among other things. The *Paviljoen* was just big enough to accommodate all pioneers. They had to sleep in fours in bunk beds in small rooms. The space between the beds was so narrow that they could not stand up at the same time.

The building could not be heated properly in the winter. There was a sort of central heating system, but it did not work good when the outside temperature was really cold. Moreover, the coal quickly ran out in the harsh winter of 1941–42 and the heating pipes froze. 'The frost lay on our blankets', remembered Lily Kettner, who came from Vienna and was then 18 years old. In the living space, there was a central stove, which, because of a lack of coal, was fired with insufficiently dried wood that produced a lot of smoke. The stove was also used to cook on. Again, according to Kettner, the food supply in Loosdrecht was on the meager side. There was repeatedly too little bread for breakfast. This was supplemented with barley gruel, which was also often on the menu in the evening. Luckily, most pioneers could get extra food from the farmers where they worked.

Weighing them down more heavily than these practical problems was the pressure of being alone without other family members, worrying about the rest of the family and the fear about what the future would bring. The young pioneers were often very depressed, which led to behavioral problems for some of them. They had fled from the Nazis, but the Nazis had now caught up with them. They felt that they were in a trap. The ideal of freedom in Palestine or just anywhere was far away.

Despite or just because of these difficult circumstances, a strong feeling of togetherness existed between the pioneers in Loosdrecht. 'The *Paviljoen* was not a children's' home, it was not an orphans' home, it was home, a real home ... We were one big family ...,' Lily Kettner stated after the war. According to Lily, the leaders created the 'magical power of the *Paviljoen*', 'the warmth that came from them, the fact that they gave themselves completely, that they were with us day and night, that they were our father and mother, brother and sister, teacher and counsellor. That they knew how they had to listen to us, simply listen, and because we knew that they were absolutely reliable ...'[11]

The Position of the Palestine Pioneers

After May 1940, many Palestine pioneers thought that the Germans would soon arrest them. This did not happen, however, and it was initially possible to continue pre-war life. At the end of 1940 and at the beginning of 1941, it even appeared possible to leave the Netherlands legally. This was because of the plans of Jewish Council employee Gertrude van Tijn for emigration to the Dominican Republic, also called Santo Domingo. These plans arose from the Conference of Evian (in July 1938) about the Jewish refugee problem. During the conference, the Dominican Republic was the only country to declare that it was ready to accept 100,000 Jewish refugees. However, the government wanted a hefty financial compensation for her hospitality. Very little came of this offer; only about 700 mostly German Jews were able to settle in an agrarian community. In October 1940, the *Joint* (the large American-Jewish aid organization, *Joint Distribution Committee*) told Van Tijn that 40 to 50 pioneers could go to this community. In March 1941, these plans, for which several pioneers learned Spanish, still existed.[12]

Nothing further came of the Santo Domingo plan, of which it is not clear whether the Germans intended it seriously. It did, however, lead to sharp controversies among the pioneers. Several of them, including Shushu Simon, rejected all emigration other than to Palestine. Others did not object to going to the Dominican Republic or to Palestine via that country, as long as they could escape from the occupied Netherlands, where the circumstances were becoming increasingly more threatening.[13]

This threatening atmosphere came from the gradual introduction of anti-Jewish measures, such as the announcement of the dismissal of Jewish civil servants in November 1940 and the NSB's anti-Semitic riots. When an NSB member died in a fight with Jews in Amsterdam at the beginning of February 1941, the *Sicherheitspolizei (SiPo, Security police)* arrested more than 400 Jews between

---

[11] Pinkhof and Brasz, *De Jeugdalijah*, 26.
[12] Bernard Wasserstein, *Gertrude van Tijn* (Amsterdam: Nieuw Amsterdam, 2013), 116, 117.
[13] GFHA, Cat. no. 84, Verklaring (statement) Arthur Heinrich, september 1955, 1.

Fig. 3: The roll call of the members of the Werkdorp Wieringermeer when the Germans removed most of them on 21 March 1941

20 and 35 years of age on the Waterlooplein in the center of the city.[14] They were taken to the Buchenwald concentration camp and then to Mauthausen, where all but one of them died shortly afterwards. The first death notices of deportees reached their families in March 1941. This made a deep impression in Jewish circles.

In this same month, the *Werkdorp* was closed after complaints by members of the NSB from the neighborhood. Approximately 240 pioneers were taken to Amsterdam; 60 others were allowed to stay until September to reap the harvest. As a reprisal for several acts of sabotage, 60 members of the total group were arrested in June 1941 and taken to Mauthausen with 240 other young Jews from Amsterdam. At the request of the German authorities, Gertrude van Tijn had given the *SiPo* the addresses of the pioneers. She was told that they would be allowed to return to the *Werkdorp*. Death notices of this group also arrived shortly afterwards.

---

14 The *Sicherheitspolizei* (SiPo) and the *Sicherheitsdienst* (SD) were merged in 1939. In practice, the SiPo did the rougher police work, such as arrests and the SD functioned as an information service, which mainly worked in the office.

In mid-February 1941, the Jewish Council was set up by German order as a contact organization between the Germans and the Jewish community. The chairman of the twenty-man strong Council was the diamond merchant Abraham Asscher. Prof. Dr. David Cohen, who was also in the Zionist union, was secretary and the 'practical brain' of the Council. The remaining members were mainly dignitaries, with several retailers and a representative from workers' circles.[15] The more than 20,000 Jewish refugees who came from mainly Germany were not represented.

The Palestine pioneers had ambivalent feelings about the Council. On the one hand, there was mistrust, which increased strongly after the deportation of the *Werkdorp* pioneers. On the other hand, it was clear that several leaders of the Jewish Council, led by David Cohen (the brother of Ru Cohen of the Deventer Union) were sympathetic toward their organization.

This sympathy also existed with Dr. Jakub Edelstein, who the Germans sent to Amsterdam in March 1941 to provide advice to the recently-founded Jewish Council. He had been a leader of the Jewish community in the former Czechoslovakia. Edelstein was a member of the socialist *Poale Zion* and had led the *Hechalutz* in his country. After the German occupation of his country, he had searched for emigration possibilities for Czech Jews. To help him, he had even been given permission to visit Palestine in May 1939.

At the end of 1939, Edelstein was involved in a German experiment with a Jewish colony in Nisko near Lublin in Poland. He had cooperated with this experiment because he saw it as a possibility to train people to become farmers and craftsmen for Palestine. In October 1939, 2,600 Jews from Moravia and Vienna were sent to this inhospitable region. The chaotic experiment, whereby the Germans had forced many of the people involved across the border to the Soviet Union, was ceased several months later. The remaining participants were then sent to Polish ghettoes. About 500 were allowed to return to their former homes.

Edelstein, who would probably also have heard in Nisko of the German murder of Jews after the Polish capitulation, was certain that deportations to Poland had to be prevented at all costs. To win time, the Czech Jews would have to work for the German industry in their own country. 'Jewish work to save Jewish lives' was his motto. He would also have given the same advice to the Jewish Council: win time, cooperate in setting up work camps in your own country, prevent deportations 'to the east'. Just as the leaders of this Council, Edelstein hoped for a speedy Allied invasion. He was, however, more pessimistic than Asscher and

---

15 De Jong, *Koninkrijk, dl. 5a*, 508–528.

Cohen about how the war would go and about the ruthlessness of the German anti-Jewish policy.[16]

Jakub Edelstein turned up unexpectedly in Loosdrecht in March or April 1941 for a discussion with the youth leaders. No report was kept of this discussion, which took place in Ivrit. In a post-war statement, Menachem Pinkhof, who was present, wrote that Edelstein was very pessimistic about the fate of European Jews, 'at least if the war does not end shortly, and there is absolutely no chance of this'. He also probably warned the leaders about anti-Jewish measures, such as those that had already been implemented in Czechoslovakia, and advised them not to cooperate with deportation to Poland and 'to save what can be saved'.[17] This meant taking the initiative, buying time and keeping one's distance from the Jewish Council.

German Plans with the Jews

In fact, Edelstein, who became the leader of the 'model ghetto' Theresienstadt in October 1941, was used by the Germans to mislead the Jews with his ideas about emigration and work in his own country.[18] Mass emigration was impossible in wartime for practical and political reasons. Before the invasion of the Soviet Union in June 1941, Hitler had already given the order to liquidate ideological opponents in that country, including Jews. So-called *Einsatzgruppen* (SS death squads) were used for the mass executions. They murdered communist executives, possible partisans, and also large numbers of Jewish men and gypsies. From August 1941, Jewish women and children were also murdered. In total, the *Einsatzgruppen*, often helped by local collaborators and German army soldiers, murdered between one-and-a-half and two million Jews.

In early autumn of 1941, Hitler gave the order to deport Jews from Germany and Austria to ghettoes in Poland and the Baltic states. 'Non-productive Jews' who stayed there were murdered by the use of methods such as gas vans. At some time between September and December 1941, Hitler then decided that all European Jews had to be killed.

It is not known when he took this decision or what the motives were behind it. Hitler may have decided on this course of action under the influence of the

---

16 *"Theresienstadt Ghetto," Lexikon,* Edelstein, Dr. Jakub, http://www.ghetto-theresienstadt.info/pages/e/edelsteinj.htm,, accessed 5 juni 2013. See also De Jong, *Koninkrijk, dl. 5b,* 1012–1014.
17 GFHA, Cat. no. 173, Verklaringen Pinkhof.
18 De Jong, *Koninkrijk, dl. 5b,* 1012–1014. Edelstein was executed together with his family in Auschwitz in 1944. He was accused of involvement in escapes from Theresienstadt.

military successes in the war against the Soviet Union in the autumn of 1941. The Japanese attack on Pearl Harbor on 7 December was likely a deciding factor as well. Because of existing treaties, Germany now also had to declare war on the United States. In Hitler's mind, this confirmed the prediction that the Jews would unleash a world war. They had to pay for this with their lives.

At the Wannsee conference of 20 January 1942, the fifteen highest civil servants and military officers present discussed and sanctioned the execution of Hitler's plans. However, it wasn't until the summer of 1942 that the definite decisions about the speed and execution of the mass murders were specified. These murders would mainly be perpetrated in German extermination camps in thinly-populated east Poland.[19]

Uncertainty and Discussions

Edelstein's reports, the deportations to Mauthausen in February and June 1941 and the continual introduction of new anti-Semitic measures led to extensive discussions among the Palestine pioneers in the second half of 1941 and the first months of 1942. These measures included an obligation for Jews to register where they lived, the formation of separate Jewish schools, a ban on travel and the creation of work camps for unemployed Jews. The central question here was what one should do: obey the German measures, look for ways to avoid them or simply refuse to do what the Nazis ordered? According to a post-war report by Menachem Pinkhof, 'the leaders of the pioneers generally had no clear idea about going into hiding or avoiding capture until 1942'. The policy with regard to the Palestine pioneers was the responsibility of different commissions of the Jewish Council. Pinkhof noted:[20]

> The active members of these commissions were mainly good Zionists, who regarded the chalutzim as a very positive element. Afterwards, when the transports started, they did everything possible to keep the chalutzim in Holland as long as possible. Going into hiding was the subject of hefty discussions in the Hechalutz. It was known that there were smuggling routes to the neutral outside world. Yet this was something that was only possible for a few people and then only for people with lots of money.

---

19 Christopher Browning, *The Origins of the Final Solution* (Lincoln: Nebraska University Press, 2004), 309–416, mainly 370–373. See also Peter Longerich, *Politik der Vernichtung* (Münich: Piper, 1998), 440–456. Bart van der Boom, *'Wij weten niets van hun lot'* (Amsterdam: Boom, 2012), 32–42.
20 GFHA, Cat. no. 173, Verklaringen Pinkhof.

The level of demoralization among the young pioneers was also an important factor in the discussions about whether to cooperate with the German measures. They were often unemployed for months, had no income and no connections in Amsterdam. Transport to Westerbork would in any case put an end to this situation.

The uncertainty of their situation was clearly described by Alice Tischauer, who was born in Breslau in 1920. She was a member of *Habonim* from 1933 and went to *hachshara* near Cologne four years later. In July 1939, Alice fled to the Netherlands, after which she found work in the *Werkdorp*. After it was closed in March 1941, the Germans took her to Amsterdam. Here, Alice had contact with a group of pioneers, who were called to report for transport to Westerbork at the beginning of July 1942.

There was a period of about ten days between the call and the transport. In this time, there was much discussion among themselves and also with third parties about whether they should report for transport to Westerbork.

> Several prominent Jews among our acquaintances and caregivers thought that our chaverim should go, to give a good example to other helpless and unsupported Jews. After all, everyone had to go sooner or later. Many of my chaverim were influenced by this thought; after all, the Hechalutz was not yet active in any illegal activities at the time.[21]

On 14 July 1942, a train with 1,132 mainly foreign Jews, including several pioneers, left from Amsterdam to Westerbork. The next day, this transport was almost entirely sent to Auschwitz. This was the first of the 93 trains that left from Westerbork to the east.

In Loosdrecht, some people also initially considered cooperating with the German plans. Arthur Heinrich, who had come to the Netherlands at the end of 1938, was a pioneer born in Frankfurt am Main in 1926. He remembered after the war that rucksacks were ready for all inhabitants of the *Paviljoen* containing 'medical supplies and vitamin pills, etc.' at the end of spring 1942.[22]

However, people gradually developed a spirit of resistance, and refused to accept the German measures. Initially, this concerned the actions of pioneers who left alone or in a small group for Belgium or France, mostly with the intention of going from there to Switzerland.

---

21 GFHA, Cat. no. 238, Verklaring Alice Tischauer, april 1956, 2. By these prominent people, she probably meant Ru Cohen and the lawyer Mr. Abel Herzberg. Herzberg was involved in the leadership of the *Werkdorp Wieringermeer* for some time.
22 GFHA, Cat. no. 84, Verklaring Arthur Heinrich, september 1955, 2.

Although precise numbers are not known, it is likely that dozens of pioneers took this step—many of whom did not reach their destination. Schraga Engel for example, a pioneer of the Deventer group, born in Czechoslovakia in 1917, got no further than Paris, where he found a place to hide in the Jewish borough of the Marais. After being betrayed, he was arrested and deported to Poland, where he survived Auschwitz. Rolf Schloss, the *hachshara* youth leader in Gouda, did manage to escape to Switzerland at the beginning of 1942.[23] Most refugees appear to have come from eastern-European countries, where people were suspicious of all government organizations, including the Jewish Council; some were wealthier Dutch pioneers.

Going into Hiding: The First Contact with Westerweel

At the beginning of 1942, Shushu Simon let it be known that several pioneers from the *Werkdorp* had succeeded in fleeing to Switzerland. They came from the mostly Polish Jewish community in Scheveningen near The Hague and disappeared without telling the other pioneers. Simon said about their action: 'They are Jews from the east; they have a healthy instinct'.[24]

The stories about flight from the Netherlands formed a source of inspiration for others. In Loosdrecht, three young pioneers (brothers Arthur and Joseph Heinrich and Dov Aschheim) sought contact with the deputy mayor. They asked how they could get an identification card without a J in it. He advised them 'to lose' their identification cards, and then they would get a temporary document without a J. However, they could not travel with it. Two other pioneers set up a hiding place in an abandoned shed. When the *Paviljoen's* leaders found out about this, however, they told the people involved that they were working on a collective escape plan.[25]

This plan had been difficult to realize. According to a post-war statement by Erika Blüth, the *Jeugdalijah* leader, Shushu Simon had contacted her in February 1942 about a possible escape of the Loosdrecht pioneers, likely with Switzerland as the intended destination. To what extent an attempt was made to realize this

---

23 GFHA, Cat. no. 51, Informatie Schraga Engel. This information states that Engel went to France in January 1943. This is incorrect. He was in Paris in the autumn of 1942. See also Siegel, *Locomotieven*, 58, 59. With the promise that he would take Paul and his sister to Switzerland, Schloss borrowed a large amount of money from a relation of the family Siegel. However, he left alone.
24 GFHA, Cat. no. 173, Verklaringen Pinkhof.
25 GFHA, Cat. no. 84, Verklaring Arthur Heinrich, 2. Arthur was 15 years old in mid-1942.

escape is not clear. However, we do know that this plan was later replaced by a plan to go into hiding in the Netherlands.

Co-leader Menachem Pinkhof supported Simon with his plans. Pinkhof had decided not to cooperate with the Germans. After the deportations to Mauthausen in March and June 1941, he asked a chemist he knew for cyanide, so that he would be able to commit suicide if he was arrested. The chemist had, however, refused. This did nothing to alter Pinkhof's conviction not to cooperate with the Germans.

Menachem probably met gardener Bouke Koning, who was living in Groenekan near Utrecht, in the spring of 1942. Bouke was an acquaintance of Menachem's fiancée Mirjam Waterman. Koning had suggested to Pinkhof that he should possibly go into hiding at his home some time later. Pinkhof was interested in the idea, but finally decided that he could not abandon the young pioneers of the *Paviljoen*.

The two youth leaders discussed the escape plans with Erika Blüth in the summer of 1942; the responsibility for the Loosdrecht pioneers also weighed heavily on her. During a discussion with Menachem Pinkhof, she proposed that he should accompany them if they were called up to go to Westerbork. Shushu Simon, who had been in a concentration camp for some time in 1938, could then go into hiding.[26]

Simon and Pinkhof were able to convince their colleague Lodi Cohen, who initially wanted nothing to do with an escape, to accept their point of view. After this, they discussed the plan again with Blüth; the plan now had the form of going collectively into hiding. 'Her reaction was that she could not judge the viability of the plan, but that she would support it if we thought we could take responsibility for it.' This meant that Pinkhof and Simon would have to report to the Germans after the 'hiding operation', in order to prevent reprisals. However, she dropped this last condition later.

Through her husband Kurth Blüth, a prominent member of the Jewish Council, Erika Blüth was possibly aware of the German plans to round up the members of the *Jeugdalijah*. It was agreed with her that she could pass along a coded message by telephone to say when it would be the turn of the Loosdrecht *Paviljoen*.[27]

Now that these difficulties were overcome, the leaders of the *Paviljoen* could concentrate on planning a collective hiding of the pioneers. This was far from simple, however. During discussions at the end of July and at the beginning of

---

26 GFHA, Cat. no. 173, Verklaringen Pinkhof.
27 GFHA, Cat. no. 17, Verklaring Erika Blüth-Henschel, maart 1957, 3.

August 1942, it quickly became evident that they had insufficient non-Jewish contacts to house the approximately 50 pioneers. Several farmer's sons and Utrecht students who were initially approached to help had very little success in finding addresses. 'During the day, we worked normally and when it was dark we went looking. ... In the first two weeks, we found some places, but we noticed that the work was not progressing sufficiently to give everyone a place to hide in time', Pinkhof remembered after the war. A solution for this came from Mirjam Waterman, in whose room the discussions took place.[28]

Fig. 4: Mirjam Waterman in 1943

Mirjam, born in December 1916, was educated as a teacher and lived with her parents close to the *Paviljoen*. Her father was a wealthy diamond trader from Amsterdam, who had gone to live in Loosdrecht in 1914 to start a plant nursery. However, nothing much came of this. The Waterman family, which had five children, was strongly assimilated. The children went to what was known as a progressive school in Hilversum. The youngest two later went to the *Werkplaats Kin-*

---

28 GFHA, Cat. no. 17, ibid.

dergemeenschap (Children's' community workplace) in Bilthoven. This school, with a special pedagogical approach, a discussion of which forms part of the backdrop in the next chapter, was founded in 1926 by Kees and Betty Boeke.

Through these younger sisters, Mirjam came in contact with Joop Westerweel, who had the practical leadership over the education at the *Werkplaats*. In 1939, he offered her a job as a teacher. Westerweel, a charismatic personality with good contacts in progressive circles, seemed to her to be the person who could be of help in this situation.

* * *

In Eastern Europe, the Palestine pioneer movement was part of the mainstream Zionist movement. In the Netherlands, however, it operated in the margin of the Jewish community. The pioneers, mainly of foreign origin, were often formed in the 'new style' of the Jewish youth movement. Here, independent action, taking initiative and accepting responsibility were important qualities.

The Palestine pioneers had only contacts with the heads of the official Jewish institutions in the Netherlands, including the Jewish Council, of which part of the leadership and several employees of German origin were favorable to them. The pioneers' communities formed as it were non-conformist islands within the Jewish community in the Netherlands. Moreover, several of them knew from their own experience in the *Kristallnacht* the practice of German anti-Semitism. They generally mistrusted the Nazi measures.

The above factors were of great importance in the decision made by the Loosdrecht *Paviljoen* leaders not to report for transport to Westerbork, but to go into collective hiding in the summer of 1942.

3 Joop Westerweel and the Left-Wing Radical Milieu in the 1920s and 1930s

Joop Westerweel was a man who made an impression on people. This was not through his physique, as he was on the small side, but through his personality—or perhaps better said, his appearance. A meeting with him has been described as a confrontation, a collision with a natural phenomenon. The writer Willem van Maanen, a sharp and critical observer who knew Westerweel as a teacher and who was later involved in his resistance activities, described him as someone with a great inner strength. This started with his appearance: 'a short stocky man with the hypnotic face of a prophet. A wild wreath of grey hair grew around his wide forehead; a voice like a loud bell either motivated people or rustled them. His blue-grey eyes investigated everyone to discern truth or lies'.

According to Van Maanen, Westerweel's face was marked by a 'big strong nose and a small sharp mouth, his hands were broad with heavily grooved fingers, his body was slightly bent, as if he were a taut bow that could fire its arrow at any moment. When he walked, he placed his feet flat on the ground, not the heel first and then the rest, but the whole foot, as if he stamped'.[1]

One might consider Joop Westerweel's unbending appearance to be a mirror image of his inner being. He did not easily make compromises. He saw the political and social situation in the 1920s and 1930s in the Netherlands as a challenge to which he had to give his own answer. This required self-knowledge and courage. A worldly or spiritual authority that he could call on could support the formulation of that answer to a certain degree, but could not offer a definitive answer. Westerweel saw his actions as being determined by his own unavoidable responsibility. It was obvious that this choice would not give him an easy life, but this did not appear to interest him.

The Youth of a Worldly Preacher

How does such a character come into being? Van Maanen and others refer to his youth as the son of a minister of a small church community. Johan Gerard Westerweel was born in Zutphen in 1899 into a fairly well-to-do middle-class family.

---

[1] Willem van Maanen, "Verzet zonder geweld," *Maatstaf* 12, no. 4 (april 1964), 2–8. (From now on: Van Maanen "Verzet,").

His father Samuel, who came from Zeeland, had risen from his simple origins, working himself up to be the owner of a profitable printing business. Several years after his birth, Joop Westerweel's parents joined the *Vergadering van Gelovigen*. This was the Dutch chapter of the Plymouth Brethren and was founded around 1845. The group was named after the place in England where they first met; they were also called Darbyists, after the founder, the English minister John Nelson Darby.[2]

According to the Dutch theologian and member of the movement, W.J. Ouweneel, the *Vergadering van Gelovigen* was 'a typical product of the Reveil, the European evangelical revival in the first half of the nineteenth century.' The religion of the Brethren was a resistance to the rationalism that had obtained great influence in the Protestant churches in the eighteenth century. The Brethren were inspired by the first Christian communities. A personal experience of faith, based on the authority of the Bible, was very important. There was a close bond between the members, who were involved in the services as much as possible, celebrating the Lord's Supper on a weekly basis. The supporters of the *Vergadering van Gelovigen* avoided everything that was 'worldly', such as art, sports and politics. They were active in evangelization, education and charity work.[3]

Darby and his followers expected a speedy return of Christ, and therefore attached great value to the prophecies in the *New Testament*, such as texts from the Book of Revelation and the role that the Jews played in it. For this reason, the Brethren are regarded as forerunners of present-day Christian-Zionism, which mainly has its support in evangelical circles in the United States.[4]

Admission to the *Vergadering van Gelovigen* meant a fundamental change in the family's life. Westerweel senior's company now printed only Christian brochures and the like, and the family soon became poor. Daily life was completely dominated by the new religion. The church services mostly took place in Westerweel's house. The father, an imposing man with a long grey beard and a skull-cap, led the Lord's Supper, breaking the bread and pouring the wine. Study of the Bible was given a prominent place, and on Sunday there was no playing or sports outside; later, the children were not allowed to go out on a Sunday.

This new way of living led to conflicts in the family, which, besides Joop, consisted of an older and a younger brother and a sister. Whereas the relative poverty did not bother him, the strict rules and regulations did, because Joop

---

[2] F. van Holten, "Een eenzame man in de kerk," *Reformatorisch Dagblad*, 7 december 2000, 17. See also Willem J. Ouweneel, *"De Vergadering van Gelovigen"* (Kampen: Kok, 2002), 37–42.
[3] Ouweneel, *De Vergadering*, 28–30.
[4] Ouweneel, *ibid*. See also Gary M. Bunge, "Christian Zionism, Evangelicals and Israel," https//www.christianzionism.org., accessed 24 juli 2013.

began to regard them as not essential for the Christian faith. During his puberty, he broke with the Brethren, but not with Christianity. He continued to respect the religious conviction of his parents. His father as the unbending man of God, who sought his own way as leader of the small group of believers, made a great impression on him. In a certain sense, Joop Westerweel would follow his example, perhaps unconsciously.

Indonesian Adventure

After his break with the Brethren, Westerweel had been a member of the *Sociaal-Democratische Arbeiders Partij* (SDAP, Social Democratic Workers' Party) for about two years at the end of the First World War. In his opinion, however, this party made too many concessions to capitalism. Around 1919, he sought contact with the reformed Zutphen minister Henri van den Bergh van Eijsinga (1868–1920). This strongly socially committed theologian and philosopher was a supporter of Tolstoy. Through his SDAP membership and his study of Hegelian philosophy, the minister had evolved into a revolutionary socialist. However, he had kept his Christian principles.

Before his unexpected death, Van den Bergh intended to set up a chapter of the Communist Party of Holland in Zutphen.[5] Westerweel would remain true to the synthesis of Christianity and (revolutionary) socialism, as Van den Bergh van Eijsinga explained in countless writings and at meetings, for the rest of his life. He later called the Zutphen minister, with whom he also had a personal bond, 'the great master who woke me up, who taught me to think, who taught me to love and who taught me to hate …'.[6]

In the meantime, after the Zutphen HBS (the Dutch High school at the time) Joop Westerweel had studied at a Christian Kweekschool (teachers training college) in Nijmegen. This last study was accompanied by the usual conflicts, because Joop did not agree with the pedagogical principles of the program. He therefore did not take his final examination there, but chose the state examination for headmaster, which he took in the summer of 1922 in Deventer. Two years earlier, he had married Hendrika (Riek) L.W. Kraan in Zutphen; at the time, they were both 21 years old.

---

5 H. Noordegraaf, "Bergh van Eysinga, H.W.P.E. van den," *Biografisch Woordenboek van het Socialisme en de Arbeidersbeweging (bwsa), dl. 3*, (Amsterdam: International Institute of Social History, 1988), 10–12.
6 Quoted in: Van Maanen, "Verzet," 4. Westerweel probably did not keep a diary, but now and then made notes in diaries, of which a small number have been preserved.

Several months after his graduation, 'J.G. Westerweel and wife' sailed on 7 January 1923 on the *Prinses Juliana* from Amsterdam to the Dutch East Indies, to work there as a teacher.[7] What made them decide to do this is not clear. In the Netherlands, jobs were hard to find because of economic problems in the early 1920s. Left-wing socialist Joop Westerweel possibly wanted to avoid compulsory military service or wanted to be involved in the education of Indonesians. In addition, a desire for adventure may also have been part of his motivation.[8]

It is certain that Joop Westerweel was not welcome in the East Indies. The Governor-General's cabinet, which gathered information about education personnel sent from the Netherlands, had sent a secret message to the *Ministerie van Koloniën* (Ministry of the Colonies) in The Hague at the beginning of November 1922. In this message, the arrival of Westerweel was 'most seriously' advised against because of his supposed communist views. Initially, the Minister of Education agreed, but later revised his opinion, because Westerweel, who as far as they knew had never been a member of the communist party, 'had not openly professed any communist propaganda'.[9] For the generally conservative colonialists, however, the difference between Westerweel's left-wing socialism and the communist party would have been of little importance.

On 18 December 1922, the director of the East Indies '*Departement van Onderwijs en Eredienst*' (Department of Education and Religious Affairs) then made a 'very urgent' final appeal to The Hague to prevent the arrival of Westerweel and his wife, who was 'of equal political persuasion'. He motivated his appeal with the argument that 'almost no region on Java could be found where communist principles have not found supporters to a lesser or greater degree.' There were also centers of communist activities in the '*Buitenbezittingen*' (Outer regions). At the end of December, the *Procureur-generaal bij het Hooggerechthof* (Procurator-general of the Supreme Court) of the Dutch East Indies, Wolterbeek Muller, joined in the objections against Westerweel and his wife.[10] However, the minister did not change his mind.

At the end of February 1923, Joop Westerweel was working as a teacher "second class" at the '*Europeesche lagere school*' (ELS, European primary school) in

---

7 *Indische Courant*, 2 februari 1923, 5.
8 A younger brother, who had also broken with his parents' religion, was involved in smuggling to Germany at the end of the First World War. He then made several long journeys and settled in the United States in 1930. An older brother was a member of the NSB in the 1930s.
9 Nationaal Archief in Den Haag (NA), Archief van het Ministerie van Koloniën, Brieven en telegram van de Gouverneur-Generaal aan het Ministerie van Koloniën, december 1922, Mail Rapportenlijst, Geheim, 2.10.36.04, inventarisnr. 2719.
10 Ibid.

Tandjong Karong in the traditionally troublesome Lampongs district on South Sumatra. This was not a pleasant stay. Joop and his wife, both total abstainers, did not mix in well with other Dutch expats. In addition, there were personal problems and health complaints. Moreover, the Indonesian political police service kept a close eye on Westerweel, whereby 'illegal activities' were ascertained (probably contacts with Indonesian nationalists).[11]

In September 1924, Westerweel was transferred to Batavia, where he was given a position with the 2nd ELS. On Java, Joop Westerweel could be called up for so-called militia service. This Militia was founded in 1918 as a sort of auxiliary corps for the *Koninklijk Nederlands-Indische Leger, KNIL* (Royal Netherlands-East Indies army), which consisted of professional soldiers. Dutchmen and Indonesian-Dutchmen were subject to military duty, and had to serve for more than five months. The KNIL (and thus also the militia) were intended to repel foreign invaders, but also were tasked with keeping domestic order. Up to the end of 1939, the compulsory military service applied only on Java and in several oil-rich districts outside Java.[12]

It will hardly be surprising that Westerweel, who had developed an aversion to the colonial system in the East Indies, refused to serve. 'The Militia', he noted in his diaries, 'would certainly be even more contemptible (than the colonial system it serves, hs). What had to be upheld by such means must be thoroughly rotten.'[13] Westerweel, possibly the first conscientious objector in the East Indies, was interned for a few months and then expelled to the Netherlands.

Although during this period the colonial press regularly wrote extensively about questions of refusal of military service in the Netherlands, nothing was published about the matter. However, the press silence was not unusual. The East Indies government usually resolved this sort of question in private, especially when (as in this case) a conflict with The Hague was possible. Because of his health problems, Joop Westerweel was exempted from military service and educational work in the East Indies. On 25 January 1925, the department stated that 'teacher 2nd class J.G. Westerweel in Batavia was honorably discharged from the country's service at his request, because of physical unsuitability'. Joop Wester-

---

11 *De Sumatra Post*, 21 februari 1923, 7. Westerweel, Johan Gerard, in: Door CID opgestelde lijst van links-extremistische personen geordend per gemeente, met alfabetische klapper, 1939, https//www.historici.nl. accessed 4 februari 2014. Here, Westerweel was described as a: 'conscientious objector; in the matter of the illegal action of fired teacher in the Dutch East Indies.' Joop Westerweel probably acted as an individual. There are no indications that he was linked to a left-wing organisation.
12 *Algemeen Handelsblad*, 20 september 1924, 8. See also De Jong, *Koninkrijk, dl. 11a*, 623, 624.
13 Van Maanen, quote, "Verzet," 5.

weel and his wife had already set sail on the *Slamat* for Rotterdam at the beginning of January.[14]

An Anarchist and Pacifist

Once he was back in the Netherlands, Joop Westerweel, who had become radicalized through his experiences in the East Indies, became a member of generally extreme left-wing organizations. At the beginning of the 1960s, the anarchist Bram Storm recalled that he met him in 1925 at meetings of the Bond van Religieuze Anarcho-Communisten (BRAC, Union of Religious Anarchic Communists). This movement arose after the First World War from a merger of several other left-wing organizations. In this period, Westerweel sought contact with the 'most radical and revolutionary anti-military factions in the Netherlands …' He seemed to feel most at home with the BRAC, which had no more than 100 mainly higher educated members. Storm analyzed this choice as follows: 'The ideology of this organization best suited his own worldview, in which the significance of religion was recognized in the socialist struggle'.[15]

Westerweel started to work for the newspaper *De Vrije Communist (The Free Communist)*, later called *Bevrijding* (*Liberation*) of the BRAC. The first articles that he wrote were about the struggle of the union of Indonesian students *Perhimpoenan Indonesia* in the East Indies and the Netherlands. Their motto: 'Indonesia free from Holland' had his full agreement. According to Storm, Westerweel also made unsuccessful attempts to contact nationalistic Indonesian students who were studying in the Netherlands at the time.

As a result of several attacks by nationalists on Java in 1925, Westerweel returned to the question of colonialism in *Bevrijding*. In Indonesia, this had only led to 'barbarism'; he referred here not to the indigenous peoples, saying, 'we are the true barbarians'. Every attempt to end colonialism should be applauded. According to Westerweel, the 'indigenous rebels' must therefore be greeted as brothers. In the December issue of the same year, he turned against moderate pacifists, who only preached disarmament. That was 'untruthful bourgeois rubbish, which has no other purpose than to mislead the proletariat and keep them from their actual struggle.' Everyone who really wants peace must choose revolution, Westerweel wrote.

---

14 *Het Vaderland*, 25 januari 1925, 12. See also *De Indische Courant*, 6 januari 1925, 7.
15 Bram Storm, "Droom en daad," in *Verzet zonder geweld*, ed. Willie Westerweel et al. (self-published, 1964), 35, 36.

According to Storm, Westerweel had the reputation in the BRAC of belonging to the 'implacables', who strived for an uncompromising anarchistic socialism and Christianity. Westerweel was not afraid of the consequences of his conviction. For example, he refused to pay taxes, because part of them would be used for military purposes. This led at least once to the seizure and sale of his possessions. He also never allowed an identity document into his house. In a time of compulsory voter turnout for elections, he systematically tore up valid polling cards. He also insisted that others share in this refusal to compromise. 'All or nothing' could have been his motto, whereby, according to Storm, he had no eye for the 'human shortcomings' of others.[16]

Joop Westerweel would not stay active in the BRAC for long. In the February 1926 issue of *Bevrijding*, there was a small argument with chairman A.R. de Jong about whether members of the organization were obliged to refuse military service. Joop Westerweel obviously thought that they should, but the chairman did not want to imposesuch an obligation in an anarchistic organization. It was possible that members might have 'serious reasons' for not refusing service. Whether they still fitted in the organization would have to become evident from their further endeavors and effort. 'Westerweel and I are not so far away from each other', thought De Jong. However, the situation was apparently quite different. Besides a short article in the March 1926 issue about a railway strike, Joop Westerweel no longer published in *Bevrijding*.[17]

Betty Boeke-Cadbury, the wife of Kees Boeke, the founder of the *Werkplaats Kindergemeenschap* in Bilthoven, also gives a picture of his views and activities in the 1920s. She first met Westerweel at a demonstration against the elections that took place in their Bilthoven back gardens. The anarchists and other extreme left-wing activists regarded the elections as a sort of mass deception that had little to do with real democracy. In the demonstration, a large cardboard ballot box containing several polling cards was to be burned. The police had heard of the meeting and turned up to prevent this. However, a circle of young people danced around the box, which was then burned. 'Shots were fired in the air, people were attacked and Joop was wounded on his forehead.'[18]

---

[16] Storm, "Droom,". See also J.G. Westerweel, "Het doekje voor het bloeden," *Bevrijding*, no. 60, september 1925, 163, 164. Westerweel, "Pasifisme," no. 63, december 1925, 219.
[17] J.G. Westerweel, "Onze muziek speelt door...dans op haar maat," *Bevrijding*, no. 65, februari 1926, 235, 236. See also Westerweel, "Tramstaking Zutfen-Emmerik," no. 66, 240.
[18] Betty Boeke-Cadbury, "Joop en de Werkplaats," Westerweel, *Verzet*, 15–18.

Bart de Ligt and Pacifism

Westerweel's activities were not limited to anarchistic circles. He was also active in the radical *Internationale Anti-Militaristische Vereniging* (IAMV, International Anti-Militarist Society). Later, he visited meetings of the more moderate *Jongeren Vredes Actie* (JVA, Youth peace initiative). This movement was founded in 1924 by liberal Protestant youths, but became rather quickly dominated by all sorts of other pacifists. At the beginning of the 1930s, the JVA had more than 2,000 members and was thus one of the most important pacifist organizations in the Netherlands.

In response to the horrifying events of the First World War, pacifism experienced strong growth in the 1920s in the Netherlands. A high point was the *Petitionnement tegen de Vlootwet* (Petition against the Fleet Act) of 1923, which was signed by more than 1 million people (of the roughly 7 million population). This *Petition* contributed to the upper chamber of the Dutch parliament rejecting the Fleet Act, which proposed a significant expansion of the Royal Dutch navy.

The peace movement lost influence at the beginning of the 1930s because it could not formulate an adequate response to the emergence of the aggressive right-wing radicalism in Italy and Germany. In 1933, members of left-wing circles were forced to recognize that the German workers' movement was eradicated by the Nazis without much resistance. From the middle of the 1930s, the moderate JVA, together with organizations of other pacifists, left-wing socialists and artists, played an important part in elaborating the concept of 'spiritual resilience'.[19]

This term was introduced into pacifist circles by the Protestant theologian and minister Bart de Ligt (1883–1938). He had turned to pacifism through socialism, which he began to regard, around 1910, as 'the social shaper of the Christian principle of love'. During the First World War, de Ligt had sharply turned against militarism, regarding it as a sort of extension of capitalism.

De Ligt called for people to refuse military service and to go on strike against war, and he demanded complete disarmament. His views brought him in contact with anarchists, Marxists and left-wing socialists, with whom he protested collectively against the war. This cooperation, however, also led to conflicts with his ecclesiastic superiors and the authorities. The Minister of War forbade De Ligt, who was a Dutch-Reformed minister in Nuenen (near Eindhoven), from being in the southern provinces in 1915. A year later, he was reprimanded by the synod of the Dutch Reformed Church because of his activities. Around this

---

19 Bert Gasenbeek and Chris Hietland, *Van jeugdig pacifisme naar geestelijke weerbaarheid, de Jongeren Vredes Actie (1924–1940)* (Breda: Papieren Tijger, 2012).

time, De Ligt resigned as minister because he could no longer visit his parishioners. Later still, he broke with the church and with Christianity. In 1915, 1917 and 1921, he was imprisoned for several weeks each time because of his call to people to refuse military service.

His experiences during the First World War drove De Ligt towards anarchism. He started to regard state coercion as a source of almost all evil. One symbol of this state coercion was compulsory military service, which robbed people of their freedom and which gave them the right to kill others. De Ligt went looking for ways to fight state coercion, and arrived at the concept of 'spiritual resilience'. With this, he alluded to 'a force in people, a deeply rooted moral conviction that prevents them being able or willing to kill others. It meant relinquishing physical supremacy. A characteristic of this philosophy is that through the spiritual force that one has, one appeals to the best in those one is fighting.'[20]

In 1934, De Ligt launched an elaborate proposal for a peace initiative at a congress of the *War Resisters International* in Great Britain. The '*Strijdplan tegen Oorlog en Oorlogsvoorbereiding*' (Battle plan against War and War preparation) combined old themes such as refusal to do military service with other forms of civil disobedience. The plan was formulated according to the anarchistic principle of the individual as driving force within the peace initiatives.

The emphasis on the individual formed a point of serious criticism from JVA circles and other more moderate pacifists. While they had respect for De Ligt's work, they considered the lack of an organizational framework and the heavy responsibility placed on the individual to be impractical and doomed to failure. As an alternative, the JVA developed the 'pacifistic people's defense', which featured, for example, the general strike and other forms of collective civil disobedience.

Because of his weak health, De Ligt settled in Switzerland in 1925. From there, he corresponded with people such as Gandhi, Nehru and Einstein. However, he regularly returned to the Netherlands for lectures and contacts with kindred spirits. He also published in the BRAC newspaper *Bevrijding*, of which he became editor in 1930. Exhausted by his efforts to launch a comprehensive international peace movement, he died in 1938.[21]

It is not known if Joop Westerweel ever met Bart de Ligt personally. However, De Ligt began to play as important a role in Westerweel's life as Van den Bergh van Eijsinga had done a decade earlier. Westerweel's youngest son Bart, born in

---

20 Gasenbeek and Hietland, *ibid.*, 117, 118.
21 Herman Noordegraaf, "Ligt, Batholomeus de," *bwsa dl. 3*, 123–126. See also Gasenbeek and Hietland, *Van jeugdig pacifisme*, 117–128.

1942, was named after De Ligt. The views of the two ministers—a synthesis of socialism, anarchism and pacifism and (at least from the side of Van den Bergh van Eijsinga) an uncompromising Christian faith—were closely related and served as sources of inspiration to Westerweel.

A New Love, Education and Korfball

Joop Westerweel did not have a prominent position in any of the previously mentioned organizations after 1926. The reasons for this are not clear. After his brief career as a publicist in the BRAC newspaper *Bevrijding*, he possibly regarded himself as unsuited for work in political organizations and the compromises that were often needed. However, other events occurred in his life that replaced these political activities: a new teaching job, a new love and his passionate dedication to the Amsterdam korfball club *Blauw-Wit*.

To start with the first, after his return from the Dutch East Indies, Westerweel first worked for some time at 'special educational institutes' in Belgium. These institutes were probably boarding schools, where he could join halfway through the course. In the Netherlands, this would have been difficult. In the autumn of 1925 or in the beginning of 1926, he was appointed as a teacher in a public primary school, in the Amsterdam *Spaarndammerbuurt*. This working-class area, the building of which was completed in the 1920s, was close to the western harbor area, where many of the inhabitants earned their living.

It was in October 1927 at his new job that he met Wil (Willie) Bosdriesz, who would later become his second wife. Wil had just been employed at the school. This occurred when Joop had been separated from his wife Riek for some time. The official divorce would not be pronounced until 1930; Riek Kraan remarried a short time later in London.

Wil Bosdriesz was born in 1908 in Amsterdam into a protestant middle class family, which she described, however, as 'free of confessional and political dogmas'. After the HBS, she went to the *Kweekschool* (teacher training college), where she graduated in 1926. The meeting with Westerweel was a case of 'love at first sight. Wil wrote in her memoirs, 'After a few days, I was completely under Joop's spell'. The new love led to a conflict with her parents, who saw nothing in the married teacher who was also almost ten years older than their daughter. Their resistance caused a break in their relationship with Joop, which was only restored after more than three years, in the spring of 1931.[22]

At this time, Wil Bosdriesz worked as governess for the Jewish family Neumann, and went abroad with them regularly. She was also active in the *Jongeren Vredes Actie* and had developed into an independent woman with a strong character. With regard to left-wing ideals, she was no less fervent than her boyfriend. After a short time, the pair started to live together in a very simply furnished attic room. 'The bed and the trunk that served as a table when we ate were our only furniture. We had windows that opened out onto the wide roof-gutter, giving us a lot of light and a clear and distant view. As we sat there in the evening on the windowsill with our feet in the gutter, the world seemed very good and we were unassailable.'[23]

In the meantime, Joop Westerweel had started his job in the Spaarndammerbuurt by developing his own educational concept. He approached each pupil individually and tried to determine his or her strong and weak points. He was convinced that there was always a subject that a child was interested in. He then tried to stimulate this interest with great enthusiasm. His approach and effort were highly appreciated by the education inspectorate. However, his lack of attention for the daily bureaucratic details of educational work caused him quite a few problems with this same inspectorate.

Much of Westerweel's free time in this Amsterdam period was taken up by his efforts for the *Blauw-Wit* korfball club. He served first as a board member for several years and then as the club's chairman in 1927. *Blauw-Wit* was founded by a total abstainers association, from which it had broken free in 1921. Problems in this initial period included the very low number of members and the lack of a good sports field. In the period of 1923–24, the number of members increased and several sporting successes were achieved.

The club did not start to really flourish, however, until Joop Westerweel became its chairman. Joop appeared to see in sports the possibility of developing himself outside traditional education. In September 1927, when he had only been chairman for a short time, he contacted the playground association of the borough of Plan West. This association had a large terrain in the Van Speijkstraat where *Blauw-Wit* was given the space to lay several fields. Here, the club grew into a full-fledged working-class club, with more than 250 members. In the clubhouse, the members played cards, chess and other games. Because there was not

---

22 Persoonlijk archief B. Westerweel, Westerweel, Lijn of cirkel, 33–38. This book contains the unpublished memoirs of Wil Westerweel, intended for her children and grandchildren.
23 Ibid., 39–45.

much else to do nearby, children could always be found in the fields. With important games, several thousands of people stood along the lines, watching the various events.²⁴

Fig. 5: Wil and Joop Westerweel with two of their children in the 1930s

---

24 Hans Luiten, "*De clubgeschiedenis van Blauw-Wit 1916–1941*" (BA thesis, Universiteit van Amsterdam (UVA), 2000).

Joop Westerweel did not take his task as chairman lightly. He saw that korfball had a certain civilizing power, and that it might be used to form characters. Players who, in the eyes of the chairman, were unsportsmanlike or otherwise behaved incorrectly, were spoken to by him in no uncertain terms. If there were too few players on Sunday morning, he 'opened his big mouth in the Van Speijkstraat and shouted: 'hey—game time!'. Then the bedroom curtains opened everywhere and the team was complete in a few minutes', a *Blauw-Wit* member remembered years later.

When members had financial problems, Westerweel tried to find solutions, so that their membership was not in any danger. For Westerweel, the most important elements of the sport were the ensuing spirit of cooperation and 'unity-concept'. To promote this concept, he organized five-day *Blauw-Wit* camps during Pentacost, where working-class children could enjoy a few days in the countryside. In the evenings, there was a campfire and community singing. For many of these children, these were events that they would remember for the rest of their lives.[25]

Westerweel remained chairman up to the end of 1932 or the middle of 1933, and *Blauw-Wit* became champion of the Netherlands in 1933. Then, he left with Wil to go to the *Werkplaats Kindergemeenschap* in Bilthoven, which was led by Kees and Betty Boeke-Cadbury. As was mentioned before he had previously been in contact with the Boekes in a voting rights demonstration.

Kees Boeke and the *Werkplaats*

The *Werkplaats* was founded in 1926 by Kees Boeke (1884–1966), an engineer who was educated at Delft University, and came from a liberal-protestant milieu. During his graduation research in England, Boeke came into contact with the *Quakers*, a Protestant movement that was founded in the seventeenth century and followed the example of the religious life of the first Christians. The supporters lived quietly and soberly, did not drink any alcohol and refused to enter military service. In Great Britain, the *Quakers* played an important role in various business sectors, such as banking and insurance (Barclays, Lloyd) and the manufacture of sweets (Rowntree), chocolate (Cadbury) and shoes (Clark). Boeke's

---

25 Jaap Stam, "Maatschappelijke vorming door gemengd sporten," *Volkskrant*, 8 november 2010, 6. See also Cor van Dijk, " Joop Westerweel, " *Blauw-Wit Vijftig*. Jubilee volume, januari 1967, 7–8.

wife Beatrice (Betty) Cadbury came from a family of very wealthy but socially aware chocolate manufacturers.

The couple married in 1911 and lived first for a while in Syria, where Kees worked as a teacher and inspector at *Quaker* schools. The Boeke family had to leave Syria because of the First World War. Syria belonged to the Ottoman Empire, an ally of Germany and opponent of Great Britain. They returned to England, where Kees, after a short training course, went to work as a teacher at a secondary school. He was quickly dismissed, however, because he preached his pacifistic views in the class. Kees and Betty then devoted themselves completely to opposing the war. After Kees had called upon miners in Wales to stop working for the war effort, he was arrested, given a prison sentence of six weeks and deported to the Netherlands in 1917.

Kees and Betty settled in Bilthoven, near the city of Utrecht, where they continued striving for disarmament and social change by organizing conferences and study meetings. Kees founded numerous committees, working groups and the like, in order to promote his ideals. However, in these eminently individualistic circles his efforts at coordination had little success. Most of his initiatives quickly died a soft death.

Boeke's ideas gradually evolved in an anarchistic direction. He wanted to form a community that was separate from the organization and power of the state—where decision-making would take place not through a majority of votes, but by collective agreement. As a consequence of this development in his ideas, Boeke decided (at the beginning of the 1920s) to no longer obey the laws of the state. He stopped paying taxes and fines, with the consequence that his personal possessions were sold several times and finally his house as well. Both Kees and Betty repeatedly had to serve short prison sentences for preaching in the street without permission, for example.

Kees Boeke also no longer used a passport or government services such as post and public transport. His final act was to stop using money; after 'some hesitation', his wife agreed with this. The Boeke family, which now had eight children, was able to maintain this way of living through the support of (wealthy) friends and mainly through the existence of the *Boeke Trust*. This trust contained Betty's shares in Cadbury, under the supervision of a workers' council. The trust provided regular subsidies for projects undertaken by the couple. These payouts were made anonymously, whereby the Boekes were left to believe that the gifts came from sympathizers. This was something that they were happy to believe, although they probably knew better.[26]

---

26 Julia Boeke and Cees Smit, *Inventaris Archief Werkplaats Kindergemeenschap (Bilthoven)* In-

Relinquishing the use of money led to another activity, for which they actually became best known: setting up an educational institution. This came about in 1925 because of practical financial problems with the Bilthoven Montessori School that the family's children attended. The funding of the school went through a council tax, which was obviously not paid. The children were compelled to leave the school. Kees Boeke, who had always been interested in education and raising children, but up till then with the purpose of creating a 'new world', grabbed this opportunity to take his children's education into his own hands. Children of kindred spirits and friends followed. Within several years—after all sorts of conflicts with the authorities, and with the support of friends and mainly of the *Boeke Trust*—a small school was established that continued as the *Werkplaats Kindergemeenschap*.

Boeke explained the reason for choosing precisely this name by saying that it was not his intention to 'school' children, 'but to give them the opportunity of developing through all sorts of work, each according to their own aptitude'. He wanted to set up 'a place to work'. 'The pupils at the institution were therefore called 'workers' and the teachers 'co-workers'. With projects, the 'workers' had the lead and the 'co-workes' assisted. All decision-making at the *Werkplaats* was done through consultation and agreement.

The education at 'Boeke's school' (as the *Werkplaats* was often called in defiance of the anti-scholastic character) evolved through improvisation, on the basis of Kees's social vision, which was influenced by his *Quaker* views. With the benefit of hindsight, the *Werkplaats* was part of the process of renewal that was taking place in the traditional classical education, starting around 1900 with the ideas of Maria Montessori and other reformers. This development would continue to the 1960s, when many elements of the reformed educational philosophies had in the meantime been integrated in regular education.[27]

Initially, the school attracted pupils who shared Boeke's views to a greater or lesser degree. Later, there were also applications from students from other educational institutes who were seen as problem-cases; they were allowed to follow the program in Bilthoven, even if it was only because of the income they brought in. At the end of the 1930s, a fellow pupil told Hermann von der Dunk (who had just fled from Germany and whose mother was Jewish), 'If you can learn, don't

---

leiding Geschiedenis (Amsterdam: ISSG, 1990), 1–9. See also Daniela Hooghiemstra, "*De geest in dit huis is liefderijk*" (PhD diss., Universiteit van Amsterdam, 2013), 128–130.

27 Hooghiemstra, "*De geest*," ibid. See for the educational development: Q. van der Meer and H. Bergman, *Onderwijskundigen van de 20ste eeuw* (Amsterdam/Groningen: Intermediair, 1979). See also Fr. de Jong, *Onderwijskundigen*, 142–158.

go to the *Werkplaats*'. However, he had the impression that the number of 'misfits' was relatively low.²⁸

In 1928, Boeke had to report to the police because he led a school without having the requisite qualification. A Montessori school teacher with a teaching certificate then agreed to act as headmaster, which solved the problem until she left in 1932. Her departure caused some difficulties, since the growing school—with more students, a new building and no money coming in from government subsidies—could not afford to appoint a new (qualified) teacher on a fixed salary.

The crisis at the *Werkplaats* happened to coincide with the unsatisfactory work situation of Joop and Wil Westerweel. They had temporary jobs at municipal schools in working-class districts, where teaching was not always easy. Both Joop and Wil wanted to have a more intensive and active cooperation with children or young adults than was possible during school hours. Joop got the idea of writing a letter to Kees Boeke (whose work he knew), in which he and Wil offered 'to come and help at the *Werkplaats* without financial demands'. When Boeke received the letter the next morning, he immediately cycled to Amsterdam. In Joop and Wil's attic room they quickly came to an agreement about the offer. In August 1932, the Westerweels went to work in Bilthoven.²⁹

Joop and Wil Westerweel in Bilthoven

Joop Westerweel felt entirely in his element at the *Werkplaats*. Here, he could work among anti-authoritarian and pacifistic kindred spirits, He could also put into practice the ideas he had developed in the traditional education milieu regarding studying and raising children. The family's meager paycheck, approximately 45 guilders per month, was compensated by an atmosphere of togetherness and idealism. The Westerweel and Boeke families regularly ate vegetarian evening meals together, with as many ingredients as possible from the *Werkplaats's* vegetable garden. There were also more contacts with other teachers and sympathizers of the school than was usual at a traditional school.

A second compensation for the poor pay involved the cultural activities available. As a result of Kees Boeke's contacts and his (often wealthy) relations, musicians, actors and other artists provided their services for little or no com-

---

28 Boeke and Smit, *Inventaris*. See also H.W. von der Dunk, *Terugblik bij strijklicht* (Amsterdam: Bakker 2008), 240–241.
29 Westerweel, Lijn of cirkel, 45, 46. See also Hooghiemstra, "De geest," 163, 164. Boeke-Cadbury, Westerweel *Verzet*, 15.

pensation. 'If a dance group came from India, ... they came to the *Werkplaats*, if people came from Sweden who could work beautifully with reeds and wood, they came to give a demonstration in Bilthoven.' Boeke, a reasonable, albeit unorthodox conductor, organized countless music performances, whereby 'soloists from the *Amsterdam Concertgebouw* orchestra' performed for little or no payment.[30]

At Kees Boeke's request, Joop Westerweel quickly became involved with both the organization of the *Werkplaats* and the education. Together with Boeke and other employees, he created an administrative model, in which the 'Co-workers council' became the highest management level of the *Werkplaats*. The members of this council chose a *Commissie van Beheer* (Administrative Committee) that regulated the finances and prepared the budgets and accounts. Personnel members could not claim a salary, but received an allowance from the Committee. The *Stichting Kindergemeenschap* (Children's' community foundation) Bilthoven was founded at the end of 1934 and supported the *Werkplaats*.

Because the *Werkplaats* did not have any formal examinations, Westerweel worked out a test system to mark the pupils' progress. These were traditionally divided into age groups that were specified by colors. There was purple for children under eight years, blue for the age group of eight to ten, green up to 13 years, yellow up to about 16 years and orange for the group up to about 18 years. Per color group, the children had to work on corresponding color tests. By doing two tests per month, the children could complete their color card for the relevant age group.

The 'color test system' gave both 'workers' and co-workers' a standard method to determine the skills of the children. Later on, in cooperation with more or less like-minded schools from Bussum and Amersfoort, a possibility was created to get a diploma that gave access to several follow-up studies.[31]

The arrival of Joop and Wil Westerweel in Bilthoven was initially quite a shock for the workers/pupils, whose performance level had previously not been properly tested. Muis de Bouter, who was a pupil at the *Werkplaats* when the Westerweels arrived, spoke of the paradise being disturbed. She realized that she and the others had, up to that moment, spent their days playing at the *Werkplaats*. The arrival of the new teachers put an end to 'the good life': 'Here, two worlds collided with each other; our Garden of Eden was suddenly confronted with harsh reality in the person of two energetic educationalists, who stood with both feet on the ground'.

---

30 Interview Sytske de Jong-W. Westerweel-Bosdriesz, 15 oktober 1998.
31 Hooghiemstra, "De geest," 166.

The children from the lost paradise initially behaved 'spitefully', and Joop took the brunt of their displeasure. 'He was just so different, we found him simply "crazy"; he was so noisy, he talked so quickly and loudly, he could see right through you with his bright blue eyes, and he could hold your hand until you wanted to sink into the ground.' Their dissatisfaction with the new employees/ teachers was not of long duration, however. With their enthusiasm, professional knowledge and charm, Joop and Wil quickly and almost unnoticeably conquered the resistance against them. 'Whoever encountered Joop had to succumb', De Bouter remembered after the war.[32]

Westerweel's actions indeed made an impression. Hermann von der Dunk, who later became history professor at the University of Utrecht, remembered him as someone who somehow always formed the center of any group of pupils and teachers whether they were excercising, doing chores or folk dancing. You saw 'an enormously wild head of hair, a large red head on a small bent body, busily gesticulating hands, the hands with imploring fingers insistently spread out like bats: Joop'.

This little man seemed to be everywhere at once. 'He panted after the ball on the sports field as if his whole being depended on it, he lectured in discussions or at day's end against the many small miseries of daily life, against poor work or bad behavior, something that every teacher always and everywhere lectures against or should lecture against. However, with Joop it was always more than the dutiful monotonous battle against a monotonously ineradicable problem; it was an impassioned but fascinating offensive and nobody in the firing line was to be envied. He taught biology and mathematics. Now and then he taught everything, but really he taught Dutch. No, he did not teach it, he preached it as a fanatical missionary who converted a horde of heathens to the true belief'.

Gerda Loeff, another *Werkplaats* pupil, particularly remembered the enthusiasm with which Westerweel did everything. Even some years after the war, she could still see him before her as 'a goblin', running over the field with handball and korfball. The plays that he directed were a glorious experience. Loeff played the role of the cat in the play *Van den Vos Reynaerde (Reynard the fox)*, a favourite of Westerweel's, because of the rascally 'anarchistic' traits of the fox. This was 'pure pleasure', mainly because the director seemed to enjoy it most. 'As teenagers, you think, "Joop don't be so crazy". It was actually a

---

32 Muis de Bouter, "Ik sta bij hem in de schuld," Westerweel, *Verzet*, 31.

bit embarrassing that he could be so incredibly enthusiastic and so wonderfully excitable'.[33]

With these memoirs, it must be remembered that they date from after the war and were partly included in a book that appeared on the occasion of the consecration of the Joop Westerweel Wood in Israel that dates from 1964. That the arrival of the vigorous Westerweel in Bilthoven was not appreciated by everyone was evidenced by the 'spiteful' reception he at first received. Von der Dunk describes him in his memoirs of 2008 (i.e. more than forty years later) as a highly driven teacher with a great influence on mainly older pupils. He was also full of temperament, and fanatic with regard to questions of principles and worldview. Von der Dunk saw Joop Westerweel, however, also as an 'incredible actor, who put himself in the spotlight with the greatest of pleasure.'[34]

Political Activities

During the time that the Westerweels were active at the *Werkplaats* in Bilthoven, European politics took a dramatic turn for the worse. The Nazis gained power in Germany at the beginning of 1933. They then started a rearmament program and gradually introduced a number of anti-Jewish laws. Fascist Italy occupied Ethiopia three years later and the military took power in Spain in the summer of 1936. Supported by right-wing forces and with the help of Germany and Italy, they quickly booked progress during the ensuing civil war (1936–1939) against the left-wing government. Two years later, Austria was annexed by Germany, after which Hitler occupied Czechoslovakia's Sudetenland with the assent of Great-Britain and France. In 1939, the rest of Czechoslovakia followed without consultation.

As a reaction to the murder of a German diplomat in Paris by a young Polish Jew, the German government organized in November 1938 the *Kristallnacht pogrom*, which involved large-scale destruction of synagogues, homes, shops and other Jewish possessions. During the pogrom, there were more than 90 deaths and many more wounded; 30,000 Jews were incarcerated in concentration camps for a period of time, whereby several hundred died through mistreatment. In September 1939, Germany invaded Poland and the Second World War was a fact.

---

33 Hermann von der Dunk, "Hij was overal," Westerweel, *Verzet*, 19. See also Interview Sytske De Jong-G. Loeff, 26 oktober, 1998.
34 Von der Dunk, *Terugblik*, 126.

Fig. 6: 'Workers' and 'helpers' of the Werkplaats in Bilthoven (ca. 1935). Kees Boeke at the top, in the middle, with Joop Westerweel to his left, in a dark suit. Philip Rümke is sitting three rows below Kees with his left arm in front of him. To the right of the flag we see Betty Boeke. Her daughter Candia, with a white collar is sitting in a row below her.

Considering his political interests and opinions, Joop Westerweel would certainly have followed these events. However, he and his wife initially appeared to be mainly involved with other matters. These were the new educational and management tasks, cultural activities and the daily concerns of their family, which now included three children, Leo, Ruth and Martha. During the Spanish Civil War, Joop was active in committees for Spanish refugees. This war caused a problem for the pacifists, because support to the armed resistance of the left-wing Spanish coalition conflicted with their principles of non-violence. Help to refugees was, however, a matter that was acceptable for all parties.

Little is known about the degree of Westerweel's effort for Spanish refugees and his other political activities during this period. But we do know what happened with the initiatives organized on behalf of Jewish refugees after the *Kristallnacht*. In the Netherlands, the events in Germany were met with broad indignation. This was partly because the government kept the borders closed for Jewish refugees. With the *Werkplaats*, which harboured several Jewish pupils,

Kees Boeke composed the song: 'Open the borders.' The first verse was as follows:[35]

Open de grenzen, open ze wijd	Open the borders, open them wide
Holland sta open voor alles wat lijdt	Holland stand open for everything that suffers
Zo is het vroeger altijd geweest	This is how it has always been
Zo is van oudsher de Hollandse geest	From time immemorial, the Dutch spirit

The aid initiatives were aimed at allowing Jewish refugees, mainly children, into the Netherlands from Germany. The initiatives did not stop with the song because, inspired by Joop Westerweel, the *Werkplaats* pupils started a *'Jeugdactie'* (Youth initiative) to support these refugees financially. In Bilthoven and Utrecht, there were house-to-house and street-corner collections, and folders were distributed. Together with other schools and institutions, the *Werkplaats* organized meetings to provide information and to protest.

There is a report about such a meeting in Rotterdam, written by Jan Smit, a young unemployed bricklayer. As a result of a newspaper advertisement about the meeting at the Volkshuis (community center) in the Rotterdam borough of Tuindorp Vreeswijk, someone said to him, 'one of us has to go there'. Thus he landed up in a small hall 'with about 20 young people like myself' (i.e. from workers' circles), 'and then Joop Westerweel arrived and told us what was happening at the *Werkplaats*'. Smit knew almost nothing about the *Werkplaats*, but he immediately felt that Westerweel was a person who had something to say, and who was capable of influencing young people in a 'very special manner' and making them enthusiastic.[36]

For Kees Boeke the successful 'Youth initiative' formed the motivation for a new attempt to gather various organizations together under his leadership. He thus attracted the attention of the Centrale Informatiedienst (CID, Central Information service), which had been following the Bilthoven activities to help German refugees for some time. The main motive for this attention was that Boeke was suspected of being in contact with the left-wing opposition to German National Socialism. At the end of 1930s, the Dutch government made a great effort to maintain good relations with the German government. Moreover, many Dutch officials had no sympathy with (suspected) left-wing opponents.

---

35 Hooghiemstra, *"De geest,"* 204.
36 Nederlands Instituut Oorlogs- Holocaust- en Genocide Studies (from now on NIOD) Coll. Palestina pioniers, Doc. II-296 A (A + B). Interview Mirjam Pinkhof Waterman-Jan Smit, 19 augustus, 1990, 1–10. (From now on: NIOD, Interview Pinkhof-Smit).

The CID, which was aware of Boeke's activities and of the ins and outs of the *Werkplaats* through a source located in its vicinity, published an extensive report about the *Werkplaats* in October 1938. The Amsterdam public prosecutor A. baron van Harinxma thoe Slooten had concluded that the *Werkplaats* was a 'breeding ground of revolutionaries', which should certainly not be given any co-operation.[37]

The CID reported in the summer of 1939 in an Overview that 'the well-known Kees Boeke in Bilthoven' had founded the *Landelijk Comité van de Jeugdactie Hulp aan Noodlijdenden* (National Committee of the youth initiative for help to the needy), with J.G. Westerweel as secretary. The Committee's original intention was to offer help to Jewish children who had fled from Germany.

Under the motto 'Youth helps Youth', the leaders of the National Committee wanted to expand the initiative to support and aid other needy children. About 25 youth organizations joined the initiative. In addition to Catholic and Jewish youth organizations, these also included, according to a CID report, '*radicale pacifistische* (radical pacifistists), the *Vrijzinnige Democratische Jongeren Organisatie* (liberal democratic youth organization), *the Bond van Abstinent-studeerenden* (union of abstinent students), the *Jongelieden Geheel-Onthouders Bond* (young people's total abstainers union) and the communistische Nederlandse Jeugdfederatie' (communist Dutch youth federation).' Of all these youth associations, only the latter was underlined in the report, because it was regarded as dangerous.

The National Committee had plans to become the Dutch chapter of the *World Youth Congress Movement (WYCM)*, which was founded in Geneva in 1935 and undertook activities against warfare and fascism. The focuses of this movement were the Spanish Civil war and a boycott of Japan protesting its aggression against China.

However, from a political and social perspective, the Committee set up by Boeke and Westerweel formed quite a motley collection of people. The CID's Overview therefore reported that through 'all sorts of opposition' a congress planned for the autumn would probably not take place. In its place, Kees Boeke wanted to organize a support initiative for China, aptly named the 'Week for China'.[38]

It is not known whether the China initiative took place. At the end of November 1939, the Soviet Union invaded Finland and Joop Westerweel organized an

---

[37] Hooghiemstra, "*De geest*," 204.
[38] Centrale Inlichtingen Dienst (CID), Geheim overzicht jaargang 1939, 10–19, https//www.historici.nl., accessed 17 december 2013.

aid initiative for Finnish refugees. The communist youth organization, which stood squarely behind Moscow in the war, would certainly have refused to join in. Little more was heard after this of the 'Youth Initiative's' National Committee.

The initiative did have a local follow-up, however. During a bike ride with a few friends in Den Dolder, close to Bilthoven, Philip Rümke, a pupil of the *Werkplaats*, saw several children who were clearly bored. When he spoke to them, it appeared that they were Jewish refugees from Germany, who were staying in a holiday accommodation of the Jewish Orphans' home in Utrecht. They were not allowed off the grounds, however, and could therefore not go to school. Rümke discussed this at the *Werkplaats* and proposed helping them. At the beginning of 1939, Westerweel and Boeke, together with the Zeist Refugees committee, set up a temporary school for the children in an empty Bilthoven mansion. The *Werkplaats* teachers went there to teach the children.

After warnings from the CID, however, that the *Werkplaats* might be a conduit for communist agitation, all official cooperation was refused. The Jewish asylum committee drew the conclusion that it was better to cease the cooperation with Boeke. The school was closed six months later.[39]

Conflicts in Bilthoven and a New Start

After a somewhat difficult initial period, Wil and Joop Westerweel were completely accepted and appreciated at the *Werkplaats*. Moreover, Joop had developed into the driving force of the *Werkplaats*, having given the ambitious ideals of Boeke and his followers a practical educational basis. With the 'color test system', the workers' progress in the field of education could now for the first time be measured properly. With the special diploma, pupils could start follow-up studies.

Additionally, Joop Westerweel was an easily approachable teacher for the 'workers'. He was a father figure who was always there for them. This was in contrast to the 'spiritual' Kees Boeke who, as Von der Dunk describes in his memoirs, 'floated at a certain distance from the children and spread silence when he walked solemnly through the building ...' With Joop Westerweel, there was very little spiritualization, if any. His outbursts made the ink pots vibrate in the classroom next door. For a certain group of older pupils, such as Willem van Maanen, Gerda Loeff and Philip Rümke, the energetic, sporty and non-conformist Westerweel, who was dressed mostly in a workman's donkey jacket with

---

[39] Hooghiemstra, "*De geest*," 201–206. See also Pinkhof and Brasz, *De Jeugdalijah*, 63.

belt, was a sort of hero. 'He was the person they also wanted to be', Daniela Hooghiemstra wrote in her dissertation on the *Werkplaats*.[40]

For Kees Boeke it was not easy to accept this new subordinate position. Up to the arrival of the Westerweels in 1932, he had mainly worked with women, who looked up to him and who mostly let him have his own way. In internal memos, he complained in the mid-1930s about the fact that his wishes were not always honored and that some co-workers did not take him seriously. He probably referred to Joop Westerweel, but did not mention him by name.

In the spring of 1936, the situation came to a head about money, a difficult subject for Boeke. He had gradually released the principle of a life without money, but it was unclear for the co-workers how the *Werkplaats* functioned financially. The Administrative Committee, consisting of people faithful to Boeke, regulated the distribution of the incoming funds, which consisted of school money and subsidies from the *Boeke Trust*.

Joop Westerweel mentioned this distribution as 'not reasonable' in a letter to Kees Boeke. The Boeke family was thought to benefit most from this system. When Kees wanted to have well-known musicians play in his orchestra, the teachers received less money. An amount of 700 guilders was suddenly set aside to be able to hire his daughter Helen as an employee. Mysteriously, there was always money available for Boeke's journeys to the United States, for example. Other co-workers had to manage with much less or got nothing.

The money conflict was finally settled. However, from Boeke's personal notes, it appears that he increasingly looked back with nostalgia to the pioneer days, when the 'strike force' led solely by himself had started up the *Werkplaats*. The idealism of this time had gradually dissipated and, in his opinion, had been replaced by 'the tests'. As a result, there was hardly any time left over for the broader development of the 'workers' in the form of excursions, camps and boat trips. Again, the identity of the person responsible for this state of affairs was not mentioned in the notes, but it was clear that Boeke referred to headmaster Westerweel.[41]

In short, there was a clash of personalities and there were conflicts of competence. At the end of the 1930s, Von der Dunk's mother told him that there were school meetings whereby 'Kees and Joop walked away in turn and declared excitedly that they were no longer involved in the *Werkplaats*'. These conflicts were

---

40 Hooghiemstra, "*De geest*," 164, 207.
41 Ibid., 179–181.

settled, the awareness 'of being a small, special island in the traditionalistic Netherlands' playing an important role.[42]

There was, however, another matter. Joop Westerweel was also an anarchist—or perhaps better said a libertine—in his personal life, who made his own rules in sexual matters. He was a recognized womanizer, who had the habit of physically greeting women and older girls of the *Werkplaats* by placing his hand on their backs or bottoms. In combination with his overwhelming personality, this conduct led to several affairs, lots of gossip and problems with his own marriage. In addition, Kees Boeke, a supporter of the *Rein Leven* (Pure life) culture, was a puritan in sexual matters. He and others at the *Werkplaats* wanted to have nothing to do with Westerweel's sexual behavior.

At the beginning of 1940, Wil Westerweel had had enough of her husband's behavior. There was a new affair—probably with Mirjam Waterman (who had got a job at the *Werkplaats* through Joop)—and that was, from her perspective, one too many. Joop withdrew to the *Woodbrookers* house in Barchem to reflect, and later told one of Boeke's employees in mid-March that he was leaving the *Werkplaats*. He would put an end to 'living together' with Wil. 'I want to live according to the truth of my heart, that's what I stand for. Everything else is mere haggling; I don't want it', he noted.[43] Wil went to Amsterdam, Joop went to live in Wassenaar, where he was already working as a private teacher.[44]

Whether there were some people faithful to Boeke who exercised pressure to make Joop leave is not clear. But this conclusion can be drawn from a note in Westerweel's diary. Here, he referred to the biblical text of Jeremiah 31–15: "A complaint is heard in Rama." Again, and again I think of this at this time'.[45] The complaint related to taking children away from mothers and sending the children to Babylon during the period of the exile. Apparently, Joop Westerweel thought that his children, i.e. the pupils of *Werkplaats*, had also been taken away from him.

To Rotterdam

Several days before the German army invaded the Netherlands on 10 May 1940, Joop Westerweel applied for the position of director at the Montessori school on the Beukelsdijk in the center of Rotterdam. He was accepted; after the war, chair-

---

42 Von der Dunk, *Terugblik*, 109.
43 GFHA, Cat. no. 259, Joop Westerweel, Dagaantekening, 15 maart, 1940.
44 Hooghiemstra, "*De geest*," 207–209.
45 GFHA, Cat. no. 259, Joop Westerweel, Dagaantekening, 11 april, 1940.

man of the school board A. Loeff stated that 'a short talk with him ... was sufficient to see that there was a vast difference between him and the other (applicants).' This difference related to his wide educational experience, but especially his more than seven years as principal at the *Werkplaats* made a big impression.

However, the chairman was sensible enough to warn the deputy headmaster: 'You can expect difficulties if you choose him as headmaster ...but a lot is going to happen in the school!' The condition was that Joop would follow the Montessori course in Utrecht for a year, from September 1940 to September 1941. He already knew much about the Montessori method, and as could be expected, there were often 'lively debates with the people teaching the course'.[46]

A Man Full of Contradictions?

To call Joop Westerweel a man full of contradictions would be incorrect. He was too consistent in his uncompromising socialism and the continual support he gave to oppressed groups of people. Still, there were inconsistencies in his attitude and behavior. For example, there is the East Indies episode, whereby the question remains what motivated him, as a left-wing socialist, to involve himself with the colonial system. To help the Indonesians? To spread the revolution? Westerweel never said anything about those subjects. He always placed the emphasis on his refusal to do military service and his imprisonment. These were only incidental matters that were manipulated by the East Indies government to get him out of the country.

His refusal to do military service had no serious consequences for him. Because of his state of health, Joop Westerweel was only imprisoned for a relatively short time (although any time a European spent in prison in the tropics would be tough), received a medical waiver and left. He was honorably discharged and could therefore work in the Netherlands without problems.

The relatively lenient attitude of the colonial government did not prevent the embittered Westerweel from expressing his support for the violent attacks by Indonesian nationalists in 1925 and 1926 in the *Bevrijding* newssheet. How he was able to justify this attitude to coexist with his pacifism is not clear.

His choice to work in education also raises questions. In 1932, the socialist Westerweel left the Spaarndammerbuurt, a working-class borough, to go and lead Boeke's *Werkplaats* in Bilthoven. He certainly did not do this for financial

---

[46] Persoonlijke collectie B. Westerweel. A. Loeff, "In Memoriam J. Westerweel," Herdenkingsrede, 10 maart 1946.

reasons, but because of the progressive educational surroundings. The *Werkplaats*, located in the well-to-do village of Bilthoven (where hardly any working-class people lived), had unmistakably elitist traits. One might wonder how this context fitted in with his practical socialist ideals in the field of education.

At the *Werkplaats*, Joop Westerweel came in contact with completely different circles than those in which he was raised and initially had moved. Through these relations, he was quickly offered a job at the Montessori school in Rotterdam after leaving Bilthoven. Before the war, this school system was also considered to be quite elitist. Joop Westerweel had clearly grown away from his radical past and also appears to have abandoned most contacts from this time. The jobs that he accepted as assistant of 'umbrella organizer' Kees Boeke, were rather 'civilized', in comparison. Joop had unmistakably remained the practical idealist who offered help to groups that found themselves in a tight corner, as evidenced by his efforts on behalf of Spanish, Jewish and Finnish refugees.

Then there was the anarchistically-inspired 'free love'. Although he definitely loved his wife, he caused her a great deal of sorrow and brought himself into difficulties. His almost certainly forced departure from the *Werkplaats* was a direct consequence of this. Wil and Joop reconciled later, and their son Bart was born at the end of 1942. However, one last affair ended with the birth of an illegitimate child in 1944.

Some friends regarded his behavior as an extension of his overwhelming, 'warm' personality. It indeed did not have much to do with anarchism, which placed the emphasis on personal responsibility. Later in the war, Westerweel's views about relationships and sexuality disturbed people who otherwise appreciated him greatly. This was also because they thought that this attitude could cause dangerous situations in resistance work. In 1944, at the beginning of her imprisonment in various concentration camps, his wife Wil decided that, should she survive, she would never live with Joop again, even though he remained the great love of her life to the end of her days.[47]

During the war, Joop Westerweel would nevertheless apply the same 'overwhelming' character, courage and idealism to take the Palestine pioneers working with him to safety. One could say that the popular preacher had found his own mission in this rescue operation.

---

47 Joop Westerweel, "Brieven van Joop uit de gevangenis," Westerweel, *Verzet*, 58, 59. See also NIOD, Interview Pinkhof-Smit, 2.

4 The Birth of the Westerweel Group

The previous chapter sketched the left-wing progressive milieu in which Joop Westerweel was active between WWI and WWII. Mirjam Waterman's request that Westerweel manage the hiding of *Jeugdalijah* pioneers in Loosdrecht in the summer of 1942 followed from his activities in those years. This chapter first pays some attention to Westerweel's time as head of the Rotterdam Montessori school on the Beukelsdijk and the general situation in the first years of the war.

Starting to Work in Rotterdam

Joop Westerweel initially followed the supplementary Montessori course in Utrecht, but in the second half of this course (at the beginning of 1941), he started to work in Rotterdam for parts of the week. With the start of the new school year in September 1941, he threw all his energy into the schoolwork. He started a school magazine that he and the children printed themselves, initiated projects for various age groups about 'important subjects' such as the human body, the Dutch East Indies and Diergaarde Blijdorp (Blijdorp zoo, nearby school), and organized trips, exhibitions and, as he had done at the Werkplaats, the requisite theatrical performances.

Most pupils found it marvelous. However, because there were complaints about tiredness and too much responsibility, some parents began to question whether it was not all too much of a burden. The new headmaster waved aside these objections, explaining that, in this way, the children would get used to the life awaiting them and become independent in thought and deed.[1]

Just as in the *Werkplaats*, there was initially a certain resistance to Joop Westerweel in Rotterdam. A female colleague said that 'defense' was her first reaction when faced with 'such a suggestive personality'. She very soon noted, however, that a 'fresh gale' was blowing through the school. To her own surprise she had to admit that she had achieved a much greater self-confidence and could tackle more tasks than she had ever thought she could. 'Yet, it remained very hard to work with his excessive demands', she noted.[2]

Westerweel tried to keep the war away from the school as much as possible. When he went to play with a group of children in a park in the autumn of 1941,

---

[1] Persoonlijke collectie B. Westerweel. A. Loeff, "In Memoriam J. Westerweel," Herdenkingsrede, 10 maart 1946.
[2] Greet Canter Visscher-Kolff, "Herinneringen van een Montessorileidster," Westerweel, *Verzet*, 61.

he indignantly tore up a poster with the text 'Forbidden for Jews'. But the war existed, of course. He had realized this when he was involved in helping two heavily wounded German parachutists in May 1940 in Wassenaar, where he lived. One of them, who was dying, told him proudly that he had killed several Dutch soldiers. Joop Westerweel found this both horrendous and characteristic of the pernicious influence of fascism, that these were his last thoughts.[3] Lending such assistance, ironically, resulted in his being detained for some time by the police because they thought that he was a member of the NSB.

As a result of the Nazi announcement of the dismissal of all Jewish civil servants, there was a meeting of left-wing activists, probably in November 1940. This resulted in a written call, partly formulated by Joop Westerweel, to refuse 'to work any longer where the Jews are driven out! ... Protest publicly by both words and deeds, where and as much as you can. Long live the true unity of the Dutch people!'[4] A pamphlet with similar content would be distributed on a small scale.

Besides the fact that Westerweel initially refused to procure an identification card, little is known of his other left-wing political activities from the period of May 1940 to mid-1942. This relatively neutral attitude was not remarkable. During the initial period of the occupation, the Germans were cautious in their approach and kept a low profile. Protest against the phased measures to isolate the Dutch Jews came from (mainly Protestant) ecclesiastic circles, the universities of Leiden and Delft and individuals. Even though this protest was courageous, it did not lead to any mass protest—with the important exception of the February strike in Amsterdam that was organized by the Communist Party of the Netherlands in 1941. The majority of the heads of the civil service and judicial system, the board of the Secretaries-General and the Supreme Court accepted the German instructions. There was only limited organized resistance in the period from 1940 to mid-1942.

In left-wing socialist circles, some people initially regarded the struggle of the Germans and their Axis allies against the British and their allies as an 'imperialistic war' that they wanted to steer clear of. Furthermore, many pacifists did not know how to operate in the new circumstances. The Boeke household seemed to be relieved when they heard news of the Dutch army's capitulation. 'Thank god, there is no more war', was the opinion; it was thought that the strug-

---

3 Lore (Durlacher) Zimmels, "Lore Zimmels," Westerweel, *Verzet*, 98. Zimmels places the story in Rotterdam. However, it took place in Wassenaar. When Joop Westerweel was a prisoner in Camp Vught in 1944, his friends collected several statements about his help to parachutists, see chapter 10.
4 Van Maanen, "Verzet," 3.

gle in the Netherlands was in any case finished. The activities of the *Jongeren Vredes Actie* (JVA) dwindled after May 1940. Later, several members were active in the groups distributing the resistance newspaper, *De Vonk*.[5]

In addition, Joop Westerweel had his hands full with the Montessori course and his work at school. At the end of 1940 or the beginning 1941, he reconciled with Wil and together they moved into an apartment on the Pleinweg in the southern part of Rotterdam. Picking up the daily routine once more with a family of three (later four) children, naturally also required a great deal of his energy and time.

The War Rears its Ugly Head

The February strike made a deep impression on Westerweel, who noted on 26 February 1941, 'I feel that my time is approaching. I feel that it is creeping up on me: I will not falter even if it costs my life. If I have to die, then everyone who it interests may know that I stayed at my post even in the hardest times, even if it seemed completely hopeless. Perhaps it won't get that far, but I feel that my fate is going to change. Everything in me is looking for a way out'. From his words, one might get the impression that Joop Westerweel was involved in the strike, but there is no evidence for this.

He did travel once to Amsterdam around this time, to offer help to a group of German-Jewish refugees. According to his wife Wil, he returned 'very disappointed'. 'They said that they could manage their own problems and did not need help from non-Jews.'[6] A fear of provocation or betrayal may have played a part in this rejection.

It greatly annoyed Joop Westerweel that the *Nederlandse Unie* (a political organization with a mass following that was founded in July 1940 to realise reforms and take the wind out of the sails of the NSB) and the universities did not join in the February strike. They found themselves 'too important' to risk their continued existence by protesting, he thought. According to Joop, the fear of the government ordering to close the school also played an important role in the cooperation of the removal of Jewish pupils from the Werkplaats some months later. If he had still worked there, this would not have happened, he told former colleagues in Bilthoven. 'I feel partly guilty for every time some-

---

5 Hooghiemstra, *De geest*, 210. See also Gasenbeek and Hietland, *Van jeugdig pacifisme*, 134, 135.
6 GFHA, Cat. no. 258, Interview Yigal Benjamin-Wil Westerweel, 3 januari 1990.

one is kicked if I do not at least publicly testify of my revulsion', he noted at the end of February 1941.[7]

For this testimony, he used the somewhat peculiar form of a letter to *Reichskommissar* Dr. A. Seyss-Inquart, the head of the German occupying power in the Netherlands. After eighteen members of the (Rotterdam-area) Geuzen resistance group and February strikers were executed on 13 March 1941, Joop Westerweel wrote to Seyss-Inquart several days later. In this letter, he stated 'that Germany will never give up, even if Europe were to become one great ruin'. This would result in 'a horribly barren and dehumanised' whole. He then referred to the Bible, which warned that 'all human pride and evil deeds' would finally be judged by God. This also applied to the treatment of the Jews.

The letter to Seyss-Inquart, of which it is not known with certainty if it was actually sent, fits in with the 'spiritual resilience' method that was developed by Bart de Ligt and others. This involves trying to convince the other, one's opponent, by appealing to his humanity. Westerweel would also use this tactic later during train journeys or in encounters with German soldiers, to the horror and dismay of some members of his group. In a democratic country such as the Netherlands before the war, such an attempt to convince others would not have been risky. During the Nazi dictatorship, putting the spiritual resilience into practice in this way was extremely risky if not reckless.

As was said before, it is not known if Joop Westerweel used other means to show his abhorrence of the occupation regime, although it must be said at this stage he already went quite a lot further than the vast majority of Dutch people.

It was not until July 1942 that the war entered the lives of Wil and Joop very directly. When they returned from a summer camp of the Montessori school, it appeared that their house was inhabited by the Cohens, a Jewish family of four from Amsterdam. The Cohens had gone into hiding after their son and later their daughter had ignored a call to report to the Germans. Mother Cohen had then contacted Tine Segboer, an acquaintance of hers, who had been good friends with Wil Westerweel since the time of the JVA.

Tine, who had been actively helping Jews for some time, had offered to take the family to Belgium by car. The first stage of the journey ended at the home of the Westerweels, who Tine knew to be on holiday. Because Tine was arrested shortly afterwards in another attempt to take Jews across the Belgian border, the Cohen family was stranded on the Pleinweg. Joop and Wil Westerweel found them when they returned home. Because the neighbors were aware that other people were staying in the house, Wil improvised the story that they had

---

[7] GFHA, Cat. no. 259, Joop Westerweel, Dagaantekening, 26 februari 1941.

sublet the house to acquaintances who wanted to visit family in Rotterdam. The Westerweel family went on holiday for another week looking for a furnished home in the meantime. In this way, the Cohens were able to remain at the Pleinweg for some time.[8]

Shortly after this event, Mirjam Waterman contacted Joop Westerweel to ask him for help with housing the inhabitants of the *Jeugdalijah Paviljoen* in Loosdrecht. The relatively quiet life of the Westerweels thus came to an end.

Joop Westerweel and the Jews

At the end of February 1941, Joop Westerweel sent a letter to Chaja Waterman, one of his former pupils in Bilthoven and a younger sister of Mirjam. Reacting to the pressure of the anti-Semitic measures, they had become interested in religious matters and followed Talmud lessons with a Hilversum rabbi. Chaja had written a letter to Joop to tell him about this. In his answer, he wrote that he did not find this interest for the 'real pious Jewish life completely idiotic', as she apparently had feared. At school, he had often talked about Judaism and Zionism with both Jewish and non-Jewish pupils.

He did not know where this interest came from. He wrote, however, that the fact that he came from a pious family himself, was an important part of it.

> Let me tell you something, Chaja, that so vividly came to mind when I read in your letter how that one pious day warmed your heart in your own house. I saw myself in my own house again. Every morning at breakfast, my father, who looked completely like an Old Testament teacher, took his skull-cap off. Then, we all joined hands and closed our eyes, while Father reverently said the morning prayer ... There was a prayer for each meal. At the end of the meals, when we were all quiet again, Father opened the large family bible and read us the old stories and teachings from the same Torah that you are now enjoying so much ...[9]

In the autumn of 1941, Joop Westerweel sent a second letter to Chaja. He complimented her on the fact that she had remained herself and had not started to harbor 'anti-Aryan feelings', as he called it. 'At the moment, there is every danger

---

8 GFHA, Cat. no. 29, Verklaring Vrouwtje Cohen-Zwart, date unknown. The family later went into hiding at another address. Her son Bernard, who was in the resistance and lived elsewhere, was arrested in March 1944. His father, who wanted to take him blankets, was then also arrested. They both died after deportation to Eastern Europe. Vrouwtje Cohen and her daughter Jane survived the war.
9 GFHA, Cat. no. 259, Brief J. Westerweel aan Chaja Waterman, 21 februari 1941.

that you might harbor such feelings and the anti-Semites would like to see you do just that.'

Joop had said to Wil: 'I wish that I were a Jew. What would I not give to go and work at a Jewish school now, for example. It is not that my school has been cleared of its Jewish pupils, you understand that, of course—then I could no longer work there'.

Because cinemas, theatres and swimming pools were forbidden for Jews, Joop and Wil avoided them. The consequence was that they led 'an isolated life'. 'Oh, how much I would like to find a real Jewish friend now, who would not send me back to my Aryan corner, but who really showed understanding for the problem that tortures me every day and every moment: A friend who would really accept me in the circle of the despised, where I could feel at home like a fish in water.'[10]

Both letters were primarily intended to support Chaja in the increasingly distressing situation. In emphasizing the similarities between Judaism and the family's Christian faith, Westerweel wanted to show that the two faiths resembled each other and shared qualities that she should not feel embarrassed about in any way.

The influence of his background with the *Vergadering van Gelovigen* can be recognized in Westerweel's opinion about Jews. After all, for religious reasons of their own, they were positive with regard to Judaism. The idea that one had to help 'God's people', was experienced strongly in parts of the orthodox protestant church in the Netherlands. This also partly explains the large number of Dutch reformed people who were active in helping Jews and in other forms of resistance.

Joop Westerweel's somewhat exaggerated statements about his wish to be Jewish did not have so much of a religious basis. They were mainly prompted by the desire to be part of a group that was treated unjustly—yes, even despised. During the time that he was in the Dutch East Indies and afterwards, the same sentiment had led to attempts to approach Indonesian nationalists. The somewhat clumsy offer of help to the German-Jewish refugees can be explained by the same motive.

The letters Joop wrote also clearly illustrate the isolation in which he lived in the first years of the war. During his seven years at Bilthoven, he had become estranged from the left-wing socialist milieu in which he had moved in the 1920s. Besides the short revival of autumn 1940, the more pacifistically oriented circles around Kees Boeke, apparently offered no basis to become active in the

---

10 Ibid., Brief J. Westerweel aan Chaja Waterman, oktober 1941.

resistance. Out of solidarity, Joop and Wil kept to several prohibitions that Jews were subjected to. However, attempts they made to contact them were rejected. Westerweel's relative isolation was possibly reinforced by his problematic departure from the *Werkplaats*, about which there were various rumours.[11] According to Wil Westerweel, the arrival of the Cohen family at the end of July or at the beginning August 1942 was in a way an enormous relief to Joop: now, he could finally offer the help that he had wanted to give for so long.[12]

Fig. 7: Bouke Koning at the beginning of the 1940s

Loosdrecht

Mirjam Waterman asked for assistance shortly after Joop and Wil had relinquished their home to the Cohen family. As described earlier, the leaders of the *Paviljoen* (Mirjam, Menachem Pinkhof and Shushu Simon) were unable to find sufficient addresses to hide the staff and the Palestine pioneers, numbering about 50, residing there. Therefore, almost at the same time as Westerweel was approached, several others, both Jews and non-Jews, were asked to help as well.

---

11 Von der Dunk, *Terugblik*, 145.
12 GFHA, Cat. no. 258, Interview Yigal Benjamin-Wil Westerweel.

The first of these new helpers was Bouke Koning. He was born into a socialist worker's family in Akkrum, Friesland in September 1915, where both the socialist leader Troelstra and the anarchist Domela Nieuwenhuis were highly regarded. Under the influence of the ideas of Bart de Ligt, Koning refused to enter military service in 1934, which cost him a prison sentence of ten months. Through his contact with a biological dynamic horticulturalist who worked there and who had previously been head of the vegetable and fruit garden of the *Werkplaats* for some time, Koning went to Bilthoven at the end of 1930s.

At the Werkplaats he was the head of a large vegetable garden that covered a few acres and contained several greenhouses. For his new job, Koning followed a course at the agriculture and horticulture school in 1939–1941. Here, he met Mirjam Waterman, who followed the same course in connection with her intention to go to Palestine. The outbreak of war made this impossible. Because she was Jewish, Waterman was no longer allowed to follow the course in 1941. However, Bouke Koning remained in contact with her and her boyfriend Menachem Pinkhof. In the summer of 1942, he discussed with Pinkhof the possibility of going into hiding.

Koning, who married Froukje Kramer in 1936 and lived with his family in Groenekan, near Utrecht, had always remained active in the left-wing movement and was fully aware of the fate of the Jews in Germany. He therefore did not hesitate when Waterman asked him to assist with the hiding operation.[13]

Other new helpers included Elly Waterman, Mirjam's sister, and two former *Werkplaats* pupils, Philip Rümke and Candia Boeke, the daughter of Kees. In this first phase, Jaap Lambeck and his wife Jo were also involved in the 'hiding operation'. The Lambecks, who lived opposite the *Paviljoen* in Loosdrecht, were acquaintances of the Waterman family. Lambeck, a committed social democrat, was involved in spreading the illegal newspaper *Het Parool* from the autumn of 1940. Until the February strike, he worked as a furniture maker at a subsidiary of Philips: the *Nederlandse Seintoestellen Fabriek* (NSF, Dutch transmitter factory) in Hilversum, and had played an important part in expanding the strike to Hilversum.

Lambeck was arrested after the strike and held prisoner for more than three weeks. After his release, he no longer returned to the NSF, but became fully active with *Het Parool* and other resistance work. The help given to the *chalutzim* was a sort of secondary activity for him and his wife. After a German raid on his Loosdrecht house, probably at the end of 1942, he went into hiding in

---

13 GFHA, Cat. no. 117, Interview Yigal Benjamin-Bouke and Froukje Koning, 5 mei 1990.

Amsterdam. He continued his work for *Het Parool*, but was only incidentally involved with the Palestine pioneers.[14]

Jan Smit became involved through Philip Rümke. He came in contact with Westerweel at the meeting described previously for help to Jewish refugees in Rotterdam. Smit, who was born into a socialist workers' family in 1917, was active in the *Zuider-Volkshuis*, a community center in the borough of Tuindorp Vreeswijk in Rotterdam-Zuid, the part of the city south of the river. Because he was regularly unemployed as a builder due to the economic crisis, he helped to organize the activities. The *Volkshuizen* were managed by a foundation that had links with the SDAP and its youth movement.

Sometime after the meeting in Rotterdam, Jan Smit cycled to Bilthoven to take a look at the *Werkplaats*. He met Joop Westerweel who said to him: 'You have done handicraft and we have quite a lot for you to do here, not as a teacher, but there is always something to repair—do you fancy it?' Smit worked for some time as assistant to the head of the workshop.

He returned to the *Werkplaats* by coincidence in 1941. To avoid being sent to Germany as a builder, he had taken a job as a carpenter in a holiday camp in Putten on the Veluwe. There, he met the 'greens and the yellows', i.e. the older 'workers' from Bilthoven, who were there at a 'theatre camp'. They asked Smit, whom they remembered from the *Werkplaats*, to cook for them. He agreed and worked in the kitchen for several weeks. When the 'theatre camp' finished, the *Werkplaats* leaders asked him to become head of the workshop in the school. There was only a few years age difference between Jan Smit and the older 'workers', such as Philip Rümke, with whom he got along well. Sometime later, Rümke asked him if he would like to help with the 'hiding operation' in Loosdrecht. Smit agreed.[15]

The Hiding Operation

The first time that Jan Smit attended a meeting in Loosdrecht, in the summer of 1942, he met Joop Westerweel, who at the request of the others, had become their leader. According to Smit, he was very well suited for this position. Westerweel was decisive and committed—and made it clear to everyone involved that there was no way back with the hiding operation. He insisted that Jewish members of

---

[14] NIOD, Jaap Lambeck, Typescript "Herinneringen van een voormalig verzetsleider," 1–52, mainly 45.
[15] GFHA, Cat. no. 225, ongedateerde Verklaring Jan Smit. See also NIOD, Interview Pinkhof-Smit, 3.

the core group pay no attention to the alternative that was recommended by the Jewish Council (i.e. not to go into hiding, but try to get a temporary postponement of deportation by being placed on a list).

Joop Westerweel also did all he could to visit personally all potential addresses for hiding. He looked at the situation and spoke with the people who were considering making their homes available to hide others. Often, he was successful in removing their objections. To carry out these meetings, Westerweel left on his travels as soon as school was finished for the day. He slept as best he could in the train or in station waiting rooms. He would not return to Rotterdam until the morning, where his schoolwork was waiting for him again.

Through his activities, he also succeeded in creating new élan and determination among the sometimes-despairing Palestine pioneers. According to Menachem Pinkhof, the 'hiding operation' was 'now more than just an attempt to save people. It was a goal in itself, a path on which one could fight for one's own dignity; it was resistance to the oppressor'.[16] The efforts of Westerweel and his helpers created the basis for a close bond between the Jewish and non-Jewish members and led to the formation of what would later be called the Westerweel group. This group, in varying compositions, continued to exist for almost the entire war.

Westerweel, Koning and Smit were initially not very interested in Zionism, which they saw as a variant of the nationalism that was little appreciated in left-wing socialist circles. Shortly after the February strike, Westerweel wrote in a note in his diary that he saw 'Jewish nationalism' as 'a weakness', just as were other forms of nationalism. He sometimes had the idea 'that the solidarity of the Jews in this time was too much a contemporary phenomenon instead of a deep conviction, even if it has the appearance of it'.

The criticism of Zionism did little to harm the close bond and the feelings of solidarity. Both groups recognized and valued each other's idealism, which naturally also had a socialist foundation with the Palestine pioneers and formed the basis for the cooperation.

Practical Matters

With the recruitment of the newcomers, there were more successes in finding addresses to hide during the first weeks of August. Bouke Koning collected ten addresses in Friesland. These addresses included those of his parents and several

---

16 GFHA, Cat. no. 173, Verklaringen Pinkhof, ca. 1963.

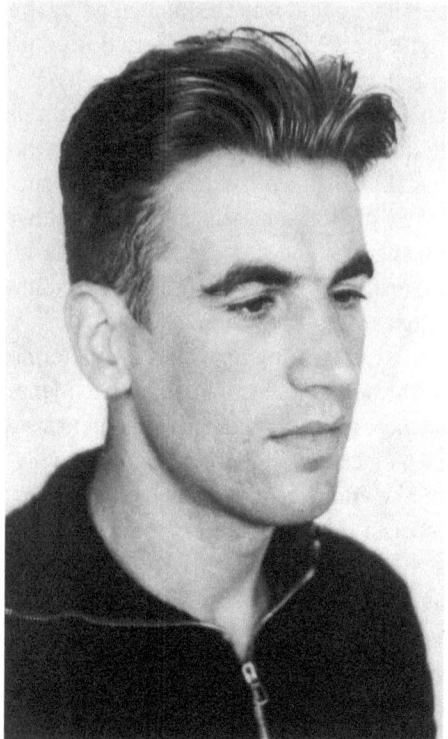

Fig. 8: Jan Smit

other family members and acquaintances. Through a friend from the AJC, the socialist youth movement, Jan Smit collected about eight addresses in IJmuiden and surroundings. Joop Westerweel was able to convince his aged mother in Apeldoorn to take a pioneer into hiding and found further addresses in Rotterdam, Amsterdam and Zutphen. Philip Rümke was able to house two pioneers with his mother, who also was active in the *Werkplaats*.

Just as important as the addresses were the false identification cards, with which the young men could travel from the *Paviljoen* to their destination and which they could use outside their hiding place. Jaap Lambeck mainly provided the false papers, which he received from two communist resistance people.

Two types of forgeries were used. With some identification cards, the 'J' was removed and replaced by a blank strip of paper with the same shape. The disadvantage of this procedure was that the forgery was fairly easy to discover when the identification card was strongly bent. The glued-in strip than often came loose from the card. With stolen or otherwise obtained identification cards of non-Jewish Dutchmen, the forgers replaced the original passport photograph with a photograph of the new owner.[17] With this forgery, the stamps on the photographs had to be reconstructed. In this case the problem was that these missing identification cards were registered.

Jews who had a somewhat dark complexion, were provided with an 'Indonesian' background. The person involved could be born there or in the Netherlands, whereas one of the parents was of Indonesian origin. For the men, this would also serve as a valid explanation for the fact that they were circumcised. Circumcision would often also be done for hygienic reasons in Indonesia with non-Islamic or non-Jewish Europeans.

Sometimes, unexpected problems occurred with the forgeries. An identification card with the 'good old' Dutch surname of Coen would serve well in the Netherlands, where it is pronounced as 'Coon'. In France, however, where the pioneer in question fled to in 1943, it was pronounced as Cohen—and that was dangerous. This man was given new papers as quickly as possible.[18]

Then, there was the question of money. During the common meetings, it was decided to pay 50 guilders a month for each person in hiding. This fairly ample amount was, in the first place, intended to cover the living expenses of the person in hiding and provide a contribution toward the extra expenses incurred by the family hiding the person or people. Sweets and possibly books and magazines also had to be paid from this money. The money to finance the first months was found by selling all food stocks in the *Paviljoen* on the black market. This yielded several thousands of guilders.[19]

A 'hiding plan' and an 'emergency plan' were now formulated. The emergency plan would be used if an evacuation order came before the whole hiding plan was ready. This was indeed what happened. On Monday 12 August 1942, *Jeugdalijah* contact Erica Blüth transmitted the code word by telephone, indicating that the Germans were going to transport all inhabitants of the Loosdrecht *Paviljoen* to Westerbork within a few days.

That evening, Menachem Pinkhof and Shushu Simon called all pioneers together in the dining hall. They told them that the leaders had decided to let them go into hiding with non-Jewish families to stay out of the hands of the Germans. 'I advised them to read and study a lot. They were very sensible. For example, some girls went to Mirjam to ask her for sexual education. Afterwards, we took each child apart and told them where they were going to and gave them in-

---

17 GFHA, Cat. no. 225, ongedateerde Verklaring Jan Smit.
18 Email Guido Smit, 23 september 2014.
19 GFHA, Cat. no. 173, Verklaringen Pinkhof.

structions.' Simon and Pinkhof also told the pioneers that they had to remember that the people who helped them were taking great risks and that they therefore had to behave well. If they were arrested, they had to say that the leaders, Simon and Pinkhof, had forced them to go into hiding.[20]

On 13 August preparations started immediately. The Jewish stars were removed from the clothing and everyone got false papers. After this, the people were taken to the addresses to hide. This was mainly done by the non-Jewish members of the group, to minimize the chances of arrest. Mirjam and Elly Waterman, who looked non-Jewish, were also involved. The pioneers were taken to their hiding addresses in the evening and night in the wide surroundings of Loosdrecht on the carrier of a bicycle. The train was used for more distant destinations. On Saturday, there were still ten inhabitants in the *Paviljoen*. However, they made quite a lot of noise to give people in neighboring houses the impression that everything was as usual.

The 'hiding operation' went flawlessly. Jan Smit experienced it as if he were in a daze, and commented that the togetherness he experienced made this time 'unforgettable'. The helpers hardly slept: 'we played music and we dug a small pit in the middle of the night, because all sorts of things had to be hidden'. He and the others had the strong feeling 'that you were doing something, that you were actively undertaking something against the Germans. We all felt that'. Joop Westerweel inspired them, saying things like, 'we must do it and we can; this is going to work. It is be possible and we are going to do it.'[21]

On Sunday morning, the last day of the operation, it appeared that there was still no accommodation for four young men. Menachem Pinkhof had his eye on an address for them in Bilthoven, but that appeared to be 'untrustworthy'. While letting the young men wait in the woods, he called Bouke Koning, who could place one of them in Friesland. Pinkhof then waited until it was dark and took the other three to an institute for children with behavioral problems in Hilversum that was run by 'Tante Kathi' Mulder. She could be trusted entirely, and had previously offered help to the young people from Loosdrecht. The three young men could stay with her for several days until other addresses were found. 'Tante Kathi' used the institute's isolation cell to hide people. Apart from herself, nobody else was allowed in this room.[22]

---

20 Ibid.
21 NIOD, Interview Pinkhof-Smit, 3.
22 GFHA, Cat. no. 173, Verklaringen Pinkhof.

On Monday morning, when the Germans appeared in Loosdrecht with several vehicles to transport the inhabitants from the *Paviljoen* to Westerbork, there was no one left.

Fig. 9: The Loosdrecht pioneers in spring 1942. Paul Sonnenberg fourth row far left, the Nussbaum sisters third row centre right. Manfred Paul with chequered shirt, third row left. Arthur Heinrich with chequered shirt, first row centre. His brother Josef with spectacles, rear row centre. Metta Lande, second row, far left. The Turteltaub brothers, second row centre, right and next to them Manfred Rübner with spectacles.

The Dangers of Being in Hiding

The 'hiding operation' was a success, but that was only the beginning of new problems. Several people had offered help had cooperated only on the condition that the pioneers in hiding would stay for a short time. This was also because some of them had a distinct, what one called, 'Jewish appearance' (lightly tinted skin, dark hair, thick eyebrows, a large nose, which was possibly bent) and spoke Dutch with a heavy German accent. The members of the Westerweel group had agreed with this condition to be able, in any case, to house everyone. Therefore, a start had to be made almost immediately with searching for new addresses.

Moreover, it became evident after just a few days how dangerous being in hiding was. Probably through Mirjam Waterman, the poet Bertus Aafjes was found willing to make his home on the Amsterdam Prinsengracht available during the two weeks that he was on holiday in Friesland. This was an attic apartment on the second floor, with space for three people. Students lived on the floor below; the owner of the building lived on the ground floor.

The three people in hiding, 17-year-old Paul Sonnenberg and the Nussbaum sisters (Gustel and Sophie), 19 and 17 years old, were taken by Shushu Simon by bicycle and train from Loosdrecht to the Prinsengracht. They were given strict instructions to do all they could to hide their presence in the apartment. This meant making no noise, flushing the toilet as little as possible and definitely not going downstairs. One helper took food and other items to them at different times. This went well for about a week, until Paul Sonnenberg decided one afternoon to investigate the house. On the first floor, he ran into a student, who warned the owner. She went upstairs, found the three of them and demanded that they leave immediately. Even when the pioneers told her that they were Jewish and that it was very dangerous to be on the street, she could not be persuaded and threatened to call the police.

Sonnenberg and the Nussbaum sisters panicked, packed some things together in a bag, forgetting their false papers, and left. It was in the early evening and they had to find somewhere to sleep before 8 o'clock. After 8 o'clock there was a curfew for Jews. The three of them walked to the Central Station, where there were many Germans in connection with the rounding up of Jews, but did not find a place to hide. Later in the evening, they decided to split up. Gustel and Sophie hid under a railway bridge and Paul went to an air-raid shelter near the station to spend the night. The air-raid shelter was known by the Germans as a place where Jews often hid during raids. Paul Sonnenberg was arrested; the sisters spent a night in fear, but were not found.[23]

The next day, Gustel remembered the Amsterdam address of Channa de Leeuw, the former leader in Loosdrecht. Her husband took Gustel and Sophie to the Palestine pioneers' house in the Tolstraat. They stayed there for two days and were then housed with the family of a socialist teacher. The son was active in the communist resistance and was sought by the Germans, so this address was not safe, either. After six weeks, Joop Westerweel, who had visited

---

23 GFHA, Cat. no. 161, Verklaring Sophie Nussbaum, december 1955, 2. In her statement, she speaks of Herengracht. This must almost certainly have been Prinsengracht.

them several times and also given them ration cards, came to take Gustel and Sophie away.

Jaap Lambeck then took them to Doorn, to the family of an instrument maker. He was friendly, but his wife 'always said that we were German. She would have preferred to have Dutch Jews'. In addition, there were problems with money. Although, according to the sisters, the Nussbaum family paid more than enough for their stay, they were not given enough to eat. After staying elsewhere for a fortnight, because of German raids nearby, Gustel and Sophie returned to Doorn for a period of three months, where a Dutch-Jewish family was also hiding in the meantime.

The tensions escalated to such an extent that the sisters had to leave. They stayed with a black-marketer for a few more weeks and were then housed separately by Joop Westerweel. Gustel returned to Amsterdam and Sophie hid in Rotterdam at a photographer's house, for whom she worked as housemaid. After wandering about again for some time, she ended up in Sevenum, in the province of Limburg. Sometime later, new problems arose at this location, this time as a result of the fighting between Germans and the Allies there in the autumn of 1944. Gustel (who was finally able to hide in Roermond) and Sophie both survived the war.

Paul Sonnenberg stayed in Westerbork until mid-1944, where he assisted with several escapes of inmates of the camp. He did not flee himself, because he wanted to stay with his father, who was also in the camp. Paul and his father, a veteran of the German army, were taken to Theresienstadt with one of the last transports. Despite all promises made by the Germans, they were transported to Auschwitz. Paul died at an unknown place during one of the death marches that took place after the camp had been cleared out.[24]

The wanderings of these three individuals (Sonnenberg and the Nussbaum sisters) have been described in some detail to illustrate the many problems that the pioneers encountered in finding a safe hiding place. One of them was their young age. Because of their relative inmaturity, some pioneers were unable to cope with the situation. Their German origin and language, their contacts with new people and habits in unfamiliar milieus and their perhaps exaggerated, but understandable mistrust, were important factors in their precarious circumstances.

In addition, not all of the people who hid them were the altruistic helpers that they were often said to have been after the war. Neither were people in hiding model guests, who bore all hardships and accepted their fate without com-

---

24 Ibid. See also GFHA, Cat. no. 229, Brief G.S. Sonnenberg aan Gustel Nussbaum, 29 augustus 1989.

plaining. Many of the younger pioneers had a hard time with their lonely stay in attic rooms. After a time, there were often tensions between both parties, which arose partly from the special circumstances and the threat of constant danger.

The wanderings of Sonnenberg and the Nussbaum sisters were certainly no exception among the pioneers from Loosdrecht. Together with his two-year-older brother Joseph, the 16-year-old Arthur Heinrich hid at seven addresses between August 1942 and April 1943. The people who helped them varied from a poverty-stricken artist and his wife, to three strict, older Catholic ladies (one of whom had worked as a missionary in Palestine), to a communist activist. When the Dutch police raided the house of the communist, the two brothers (who did not look Jewish) were found, but were released shortly afterwards. The agents were apparently more interested in arresting the entire communist family: husband, wife, daughter and son.

After a period of moving from address to address, the brothers were asked by Joop Westerweel if they were interested in going to Belgium in the spring of 1943, as the first step on the road to Spain. They agreed to go.[25]

Other Loosdrecht people in hiding had more luck with their address. Metta Lande, a 17-year-old from Vienna, stayed with a family in Zutphen for almost a year. She had to remain in her room almost permanently, which was very difficult for her. When she had to flee from Zutphen because her hiding place had been betrayed, she returned by herself for two weeks on the farm in Loosdrecht where she had worked before, until she went into hiding. Metta then went to serve as a nanny to the young family of a Remonstrant (Protestant) minister in Hilversum. There, she was allowed to go outside with the children, which she liked very much. Because Hilversum was not safe, either, she left for France in January 1944.[26]

Manfred Paul, who came from Kolberg in Pommeren, Eastern Germany, and was 17 years old in 1942, was first in hiding at several addresses in Hilversum. Joop Westerweel then took him to Rotterdam, where he stayed in a house that belonged to the brother-in-law of Tine Segboer, the resistance woman that was introduced before. This brother-in-law was a member of the NSB, but lived elsewhere temporarily. He must have known that 'guests' were staying in his house, but made no objection. The Westerweel group used the house as a *safe house* to shelter pioneers who had to leave another address. In the time that Manfred Paul lived there, the other residents included Shushu Simon and his fiancée Sophie

---

25 GFHA, Cat. no. 84, Verklaring Arthur Heinrich, september 1955.
26 GFHA, Cat. no. 121, Interview Mirjam Pinkhof-Adriaan Doets and Metta Lande, et al., 6 mei 1986.

van Coevorden (who did the housekeeping), and another three pioneers, including the 18-year-old Israel Teitelbaum.

On 19 September 1942, Paul and Teitelbaum moved to an address on the Westzeedijk, where a widow lived with her two adult sons. Two months later, they hid with a teacher's family for more than half a year. A helper of the Westerweel group took Manfred Paul and another Loosdrecht pioneer, the 19-year-old Manfred Rübner, to a guesthouse on the Heemraadsingel. This guesthouse belonged to a widow who initially did not know that her new tenants were Jews hiding from the Germans. Later, when she became aware of this, her response was to provide space to three more Jews, including a middle-aged Jewish teacher.[27]

The teacher was a difficult man, who had stayed in Arnhem previously. There, he did not adhere to the rules, and demanded extra food. Even worse was the fact that he argued with people and then threatened to go to the German authorities and betray his helpers. After a long discussion, it was decided to eliminate him. This plan did not come to pass, however. The teacher was taken to an address in Rotterdam, where he once again caused problems, after which Jan Smit and Norbert Klein (more about whom later) approached a *Knokploeg* (armed resistance group) to liquidate him. However, at the last moment Jan and Norbert said to each other, 'we can't do this. We've got to keep this guy with us and take care of him'.[28] This is how the teacher ended up on the Heemraadsingel, where he continued his dangerous behavior.

Manfred Paul was given false identity papers in the name of Marinus Pietersen from Amsterdam. After some time, Joop asked Paul if he was prepared to go to Palestine via Spain. Paul said he was and travelled to Antwerp at the beginning of July 1943. In Rotterdam, several weeks later, the Germans arrested the teacher, who liked to take long walks, after which the guesthouse was raided. Both Manfred Rübner and another pioneer, Jo Spier, were arrested.[29] A third pioneer, Ushi Levy, jumped out of a window and escaped on his bare feet. The *Sicherheitspolizei* took Rübner to Westerbork, from where he escaped shortly afterwards. However, in the autumn of 1943, he was arrested again in Rotterdam and sent to Auschwitz as a 'punishment case', where he was murdered.

Of the strictly orthodox brothers Max and Benno Turteltaub, 17 and 16 years old in 1942 respectively, Max had just as little luck. After a period of hiding in probably also the Rotterdam transition house, they were taken in by a minister's family in Wassenaar. When the Germans raided there in the summer of 1943, the

---

27 GFHA, Cat. no. 170, Verklaring Manfred Paul, oktober 1955, 2, 3.
28 NIOD, Interview Pinkhof-Smit, 4.
29 GFHA, Cat. no. 170, Verklaring Manfred Paul, 3. Jo Spier escaped later from Westerbork.

two fled to a meadow behind the house, where they hid behind a wall. After a time, Max went to look if everything was safe at the vicarage, but he walked straight into the trap of the waiting Germans. Benno watched as they took his brother away in the bright light of a searchlight.[30]

Increasing Work and Increasing Pressure

The continuing search for new hiding addresses, the always dangerous journey to these addresses and visiting pioneers in hiding to see how they were getting on meant a heavy burden for the Westerweel group. Thus, caring for people in hiding was extremely time-consuming. Wil Westerweel, who did her part of the resistance work alongside her many domestic chores, provided an overview of the work involved. First, money had to be picked up at various addresses. This money then had to be taken to addresses where pioneers were hiding, together with ration cards, books and games. These addresses were sometimes in Rotterdam and the surroundings of The Hague, but they could also be in southern province of Limburg. 'During these journeys, you could get nothing to eat in cafes anymore without ration cards, but you could often get a bowl of soup. On one day, I drank so many bowls of soup that I returned home bloated.'

Much more dangerous still were the random controls. One day, when Wil left Amsterdam with a pile of blank identification cards in a bag, all passengers at the Rotterdam station were thoroughly checked. 'The only thing I could think of was to go into an air-raid shelter. I sat there for some time and when I carefully went to check everything, the coast was free and I could go home. Only at home did I start to shake.'[31]

Other group members had similar anxious experiences.

The pressure increased even more when, in the weeks after the Loosdrecht 'hiding operation', the board of *Hechaloets Nederland* (*Hechalutz Netherlands*, the umbrella organization of Palestine pioneers of which Simon and Pinkhof were part) requested help with finding addresses to hide other pioneers. As Menachem Pinkhof wrote, going into hiding had become an act of resistance. In the first place, this was resistance against the Germans, but also against the Jewish Council, which, with its continuing compromises, was increasingly regarded by the *chalutzim* as a tool of the Germans. The urgency of their request— which Joop Westerweel and others initially felt disinclined to comply with be-

---

30 GFHA, Cat. no. 241, Aantekeningen Benno Turteltaub, ca. 1995.
31 Westerweel, Lijn of cirkel, 70.

cause it could form too heavy a load—diminished when the pioneers received a stamp at the end of September 1942. With this stamp, they were provisionally exempted from transport to Westerbork. The *Hechalutz*, however, continued making preparations for going into hiding. Several pioneers did not trust the Germans and decided to go into hiding at the beginning of 1943.

The respite from reporting for Westerbork reduced the pressure on the Westerweel group only to a limited degree. After his work at school, Joop Westerweel was almost continually on the way. He slept at home at best twice a week. However, he regarded the care given to people in hiding as a task that gave his life meaning, and so did not complain. When more money became available in 1943, Menachem Pinkhof travelled to Rotterdam to propose that Joop give up his job (the family could then receive an allowance from the *Hechalutz*, so that his life would become less hectic). As they spoke, Joop described his work at the school with such enthusiasm, however, that Pinkhof never got around to say what he had come to say.[32]

Menachem Pinkhof himself and Mirjam Waterman, both provided with false papers, along with Jan Smit, spent a great deal of time accompanying people in hiding. Bouke Koning regularly travelled to Friesland to see if everything was going well. Such tasks were not possible for Shushu Simon; due to his so-called Jewish appearance, he was mainly active behind the scenes.

From the very beginning there was thus a distribution of tasks and a work structure. The core group, composed of Joop Westerweel, Shushu Simon, Menachem Pinkhof, Jan Smit and Mirjam Waterman, often with Wil Westerweel and Bouke Koning, generally met once a week. Because of Joop's limited time, the meetings were mostly held in the attic of a small annex of the Montessori school on the Beukelsdijk. During these meetings, the people present discussed urgent and current matters, and proposals were made for improvements and solutions.

The Stress Experienced by the Helpers

The core group remained intact, but several young helpers were less able to cope with the pressure. Candia Boeke, a student at the School voor Maatschappelijk Werk (School for social work) in Amsterdam, actively helped Jewish fellow students and acquaintances from the moment the anti-Semitic measures began. After the Jews were expelled from the School, she did what she could to keep them involved in the coursework, providing false medical statements and taking

---

[32] Menachem Pinkhof, "Menachem Pinkhof," Westerweel, *Verzet*, 68–70.

food packages to detainees in the *Huis van Bewaring* (detention center). Here, the Germans subjected her several times to long interrogations.

In the early days of the 'hiding operation', as a member of the Westerweel group, she took a group of seven young men of the *Paviljoen* (five of whom spoke practically no Dutch) to Nijmegen by train. Certain prior rules were agreed on. 'I said, "we will all sit apart, we will see each other in Utrecht". There and in Arnhem, Candia had to change trains with the group. 'I said, 'make sure that you never give an answer that can put you in danger, and make sure that you behave as inconspicuously as possible; try not to talk to people, just go and sit in a corner.'[33]

When they arrived in Nijmegen, it appeared that it was a Catholic holiday. Lots of shops were closed and it was quiet in the town center. Candia had to take them through a city that neither she nor the seven young men knew, and she also had to prevent them from being seen as a group that belonged together.

> 'I said, 'keep about one hundred metres apart from each other, so that you can keep a watch on each other, so that you can see if someone crosses the road and look around to see if the person behind you has understood, so that you do not lose each other'. Well, it was terribly difficult. We arrived at the house and the people there had already told me "remember, in the house next door, there are people who should not know that you are there, so you must be deadly quiet. You may not flush the toilet, you may not do anything to make a noise, preferably no coughing or sneezing. You may stay the night there and then move on the next day." Well, the troubles I went through with those seven young men there and the one looked even more Jewish than the other.'

Candia increasingly felt: 'this is not my way, I can't do it'. She talked about her problems with Wil Westerweel, asking her for a different assignment. She was then given ration cards, newspapers and messages, which she delivered by bicycle for a time. However, Candia became increasingly nervous and finally quit her resistance work.[34]

For the 19-year-old Philip Rümke, who was busy searching for addresses and taking people into hiding, the pressure also became too much after a while. Various matters affected him, such as the depressing atmosphere in what one could call 'Jewish Amsterdam', where people he knew continually disappeared. The work finally proved too much for the youthful Rümke.

> I remember a time that I had to go from Hilversum to Utrecht by train to take away two young men, who looked so incredibly German-Jewish. The train stopped in the middle of

---

33 Interview Sytske de Jong-C. Boeke, 7 december 1998, 1–22, mainly 11–14.
34 Ibid., 14, 15.

the countryside, which was quite frightening. It later appeared that someone had pulled the emergency brake; it meant nothing. However, I was afraid that the Germans would check the train again and that the young men would be arrested, together with me. Those were anxious moments. I was not as able to cope as Jan Smit and others, but they were also all older than me. I did what I could do in the beginning but had to stop later on.

His decision was also influenced by a raid in his parents' house, in which his brother was arrested in his place. When Rümke, who was also involved in other resistance work, was called to report for compulsory work, he decided to withdraw completely from the Westerweel group at the end of 1942 or at the beginning of 1943 and go into hiding at the address of a *Werkplaats* acquaintance in Friesland.[35]

An Attempt to Escape to Switzerland

In 1941 and 1942, rumors reached the Netherlands about Jews who had managed to flee to neutral Switzerland. This was also considered in the original plans of the Loosdrecht group. However, because of the huge practical problems, this possibility was not investigated any further and the group chose to help people go into hiding in the Netherlands. But the Westerweel group quickly realized that this was a temporary solution. The heavily-populated Netherlands had few natural hiding places, there was little readiness to help Jews and there was always the danger of being betrayed. This made it very difficult to get large numbers of people into hiding. Finding an escape route to Switzerland, as had already appeared successful several times before, would relieve the pressure on the organization and offer the possibility of working with fewer risks—so reasoned the core members of the Westerweel group.

At the beginning of October 1942, Joop Westerweel probably came into contact with a group of 'people smugglers' who might be able to provide safe transport to Switzerland. How this connection came about, is not clear. According to Menachem Pinkhof, the contact was made via 'Jews, who had been smuggled across the border themselves'. Next, followed a meeting with the smugglers' agent in an Amsterdam café. In another version, there was contact through members of the NSB in Arnhem. An Amsterdam connection is, however, more likely.

The smugglers wanted 5,000 guilders per person, a very high amount for the time. For this amount, they would take the refugees through Belgium and France and over the Swiss border. The core group did speak extensively about their pro-

---

[35] Interview Sytske De Jong-Ph. Rümke, 15 oktober 1998, 1–24, mainly 10, 11.

posal. It was clear that great risks were involved in the venture. The costs were very high and it was not known how the journey was organized. But it was also clear that the situation with the people in hiding was very difficult. Joop Westerweel was 'one hundred percent convinced' of the reliability of the smugglers; his wife Wil had very strong doubts. However, Joop was able to convince the others. According to Jan Smit, who was present at the meeting, he had 'a limitless trust in everyone and everything'.[36]

Organizing the logistics of the journey took quite some doing. Partly because of the uncertainties and risks, it was decided to send the most stable and healthy of the pioneers. The informal leader of the group was Juda Pinkhof, the 21-year-old brother of Menachem. The rest were 17 and 18 years old and, apart from the Dutch Bernhard Aschheim, had German or Austrian nationality. Because neither the *chalutzim* nor the Westerweel group had the money to pay the smugglers, several possible candidates from outside the Loosdrecht group were sought who could pay both for themselves and for a pioneer. Some of them put all their money together and sold possessions to pay the travel costs.[37]

The group of sixteen people, including eight pioneers, began the journey to Switzerland around 20 October. Several days later, a letter that had been smuggled outside the prison of Brussels arrived in Rotterdam, containing the message that they had been betrayed. The people smugglers had handed the group over to the German police in Belgium.

The message reached the Westerweel group during a meeting on the Beukelsdijk. Mirjam Waterman still remembered, years later, that there was a knock on the door and that Joop went downstairs. He came back 'as white as a ghost; he then laid his head on the table and sat crying like a small child'. After a time, he pulled himself together and told us about the treachery. Immediately afterwards, however, he said that the work had to go on.[38]

More details surrounding the smuggling organization with which the Westerweel group was involved are not available. In the autumn of 1942 and at the beginning of 1943, there was a multitude of unsavory groups and individuals active in this field. Considering the large scale of the operation, the working method and Brussels as the place of arrest, it is possible that the group fell victim to the machinations of three traitors in service of the German police. The leader was a certain M.C.E.H. Weekers, a Belgian born in Limburg in 1906. The three traitors had representatives in Amsterdam, Rotterdam and The Hague, who con-

---

36 GFHA, Cat. no. 173, Verklaringen Pinkhof. See also NIOD, Interview Pinkhof-Smit, 7.
37 NIOD, ibid., 7.
38 Ibid., 6.

tacted Jews who wanted to flee to Switzerland. After the travel costs were paid, the representatives took the refugees in groups to Limburg, where they were handed over to Weekers and his helpers. They were transported to Brussels, where Weekers delivered them to the Germans.

The sixteen people who travelled via the Westerweel group, were sent from Brussels as 'punishment cases' to the barracks of Dossin near Mechelen, the Belgian Westerbork. On 31 October 1942, they were transported to Auschwitz, where they were murdered.

The practices of Weekers and his helpers came to an end in 1943, when the Belgian resistance passed on a warning about their treachery to several resistance members in the province of Limburg. From an investigation led in part by a police sergeant, it appeared that Weekers and his helpers had delivered at least 180 Jews in this way to the Germans. Weekers notebook, which had fallen into the hands of the Limburgers, clearly indicated that he had personally betrayed 150 Jews. The Amsterdam representative of the smuggling group, the former policeman H. Campers, who was probably responsible for a large part of this treachery, was liquidated by a resistance group in the centre of Amsterdam shortly afterwards. Weekers himself was condemned to death and executed after the war by the Belgian justice system.[39]

After this catastrophic failure, the Westerweel group began to search for ways to bring people in hiding to safety outside the Netherlands by themselves (see chapters 7 and 8).

* * *

The 'hiding operation' in Loosdrecht was a group activity that was clearly boosted by the efforts of Joop Westerweel and Bouke Koning. Just as the Palestine pioneers operated in the margin of Jewish Netherlands, so were the two initiators outsiders in Dutch society. Both of them had served time in prison as conscientious objectors and were therefore monitored by the Dutch police afterwards. They and their fellow initiators of the 'hiding operation', Wil Westerweel, Mirjam Waterman and Jan Smit, had also been active in the *Werkplaats* in Bilthoven. In 1940, this progressive educational institute had a separate place in the Dutch educational system and the justice department even suspected it of subversice activities.

---

39 A.P.M. Cammaert, *Het verborgen front, dl. 1* (Leeuwarden: Eisma, 1994), 441, 442. See also Jack Kooistra and Albert Oosthoek, *Recht op wraak. Liquidaties in Nederland* (Leeuwarden: PENN Uitgeverij, 2009), 172, 173.

From this position of relative social isolation, they were able to react spontaneously to the request for help from the Palestine pioneers. The people they approached to participate in the hiding operation were also often from the fringes of society. Jan Smit's AJC friends from the region of IJmuiden were simple people who had only the most basic sanitary facilities, which some of the pioneers in hiding later complained about.

The charismatic Joop Westerweel was particularly successful in motivating people from a wide range of religious or philosophical and social backgrounds to cooperate in the hiding operation. In this sense, he functioned as a sort of social crowbar in the fixed traditional patterns of Dutch society.

5 Expansion Activities and Reinforcements

Map 1: The Westerweel group: hachshara training, important hiding centres, escape routes.

The main subject of this chapter is the increase of the Westerweel group's resistance activities. After the successful hiding of the Loosdrecht group, the leaders of the *Hechalutz* umbrella organization decided to let other Palestine pioneers hide as well. To accomplish this, the gradually expanding and reinforced Westerweel group was again called on towards the end of 1942 and the beginning of 1943.

People who were considered candidates for going into hiding belonged mainly to three groups: the former inhabitants of the *Werkdorp Wieringermeer*, the Palestine pioneers in Gouda and the pioneers in Elden, near Elst in the Betuwe, the most important Dutch fruit production area. In addition, a growing number of individual pioneers from the Deventer training course and from the religious *hachshara* groups applied for help.

Starting with the former *Werkdorp* pioneers, this chapter provides an overview of the new groups of people who wanted to go into hiding.

Pioneers of the Werkdorp after its Closure

Chapter 2 describes the events surrounding the 270 inhabitants of the *Werkdorp* after its closure in March 1941. More than 60 of them stayed for six months in Wieringen to bring in the harvest. The remaining 200 or so went to Amsterdam, where they were initially given refuge in Asscher's diamond-cutting factory in the Tolstraat, in the Pijp district in Amsterdam. After a while, most of them found a place to stay with family or with Jewish private individuals. About 40 younger pioneers without families were housed in the then Plantage Franschelaan (now called the Henri Polaklaan). Others were housed in a home on the Nicolaas Witsenkade. Here, a meeting center was set up that was mainly visited by non-Zionist *Werkdorp* pioneers.

In June 1941, as a reprisal for an attack by the resistance, the Germans arrested 59 of the group who had gone to live with family or were taken in by other families. They were transported to Mauthausen and murdered shortly afterward. Fifty-two former *Werkdorp* pioneers were in the first transports, in July 1942, from Amsterdam to Westerbork, from where they were then almost immediately transported to the east. Nineteen of these pioneers were between sixteen and eighteen years old. They probably all came from the shelter on the Plantage Franschelaan.[1]

---

1 Benjamin, *They were*, 26.

During 1941, several pioneers from the *Werkdorp* participated in individual *hachshara* (agricultural training courses), mostly through the Deventer Union. About 25 pioneers formed a *kibbutz* in Enschede and went to work for various farmers. Others trained at an *Ambachtsschool* (trade school) or took part in agricultural projects in and around Amsterdam. Several female pioneers worked in education or took a course in order to become a nurse at the Jewish psychiatric hospital, *Het Apeldoornse Bos*. The *Joodse Centrale voor Beroepsopleiding* (JCB, Jewish Center for Vocational Education), which was founded in June 1940, coordinated all of these projects. To give the Zionist *Werkdorp* pioneers a place to meet each other, the Dutch *Hechalutz* board set up a center, the *Beth Chaloets* (House of the pioneers). This center was located in Amsterdam first in the Oosterparkstraat and later in the Tolstraat.[2]

Beth Chaloets as a Center of Resistance

Menachem Pinkhof and Shushu Simon were part of the *Hechalutz* management. They had kept their fellow board members informed about the progress of the Loosdrecht hiding operation. In the autumn of 1942, after the relative success of this operation, the leaders asked fellow board member Kurt Hannemann to make preparations to help other pioneers go into hiding—those from the former *Werkdorp* and from other training places. Hannemann, who was 23 years old, was chosen mainly because he was good friends with Simon. They were both from Berlin, where in the 1930s Hannemann's father owned a bookshop in the Friedrichstrasse. They had also both attended the same Zionist high school and had been active in the German *Hechalutz* movement, in the Elgut training center and elsewhere. Hannemann and Simon were arrested during the *Kristallnacht* in 1938 and were incarcerated in the Buchenwald concentration camp.[3]

Afterwards, at the beginning of 1939, Hannemann went to the Netherlands with several other pioneers, including Simon. After having worked on a farm near Assen for a while, he went to the *Werkdorp* in the Wieringermeer, where he stayed until its closure. Kurt Hannemann was a quiet young man, who came across as being shy. However, his ability to put things into perspective and his sense of humor enabled him to instill so much trust in people that he was chosen for management positions in both Elgut and the *Werkdorp*. He

---

2 Stegeman and Vorsteveld, *Het Joodse werkdorp*, 127, 158.
3 GFHA, Cat. no. 13, Kurt Benjamin, unpublished Manuscript: Das Ringen der Hechalutz, 12, 13.

lived in the *Beth Chaloets* together with Sophia van Coevorden, Gideon Drach, Lore Sieskind and others.

Fig. 10: Kurt Hannemann

Hannemann was given the task of maintaining existing contacts with non-Jewish helpers, such as the Westerweel group, and attracting other possible helpers from the same circles 'who had already helped once and who were ready to help again'. After 'Loosdrecht', the Dutch *Hechalutz* leaders had concluded that trusting only in their own strength was not enough. To be able to save a substantial number of the Palestine pioneers, they needed to collaborate with reliable non-Jewish helpers from generally left-thinking circles.

In addition to establishing contact with potential helpers, Hannemann was also responsible for raising money to fund the hiding operation as well as the day-to-day costs of those living in hiding. A large part of this money initially came from the Jewish Council. Later in the war, money was also donated by private Jewish individuals, such as Levisson, a used-textile merchant from Zutphen, and from the *Nationaal Steunfonds* (NSF, National Support Fund).[4] In addition, ration cards and stamps were needed as well as papers to provide people in hiding a new identity. For Hannemann, it was 'work that required him to be more and more resourceful and was increasingly demanding, and only too often resulted in small successes and great disappointments'. Moreover, with his pronounced Jewish appearance, he was continually in danger when he had to go somewhere without a Jewish star on his clothing.[5]

The continuing internal discussions of whether it was a good idea for the pioneers to go into hiding formed an additional burden. In these debates, an important factor that was often mentioned was the idea that the war might end fairly soon (in about two to three months). People asked why they should run the risk of being arrested while in hiding and be sent to Westerbork as a 'punishment case', when it was possible to be legally safe for a while, after having obtained a stamp in their papers entitling postponement of deportation. Another argument was that the circumstances in Poland were perhaps not as bad as had been reported. Hannemann had to convince these people that they would be better off taking their fate into their own hands, rather than waiting passively. Many pioneers had lost their self-confidence through the continuing uncertainty and fear; through his efforts, Hannemann gave them new confidence.[6]

The discussions with Kurt Benjamin, a fellow pioneer from the *Werkdorp*, were of a different nature. He felt that no time should be wasted on finding hiding places in the Netherlands. It would be better to take the people to Belgium and France as quickly as possible.

In late autumn 1942 and at the beginning of 1943, Hannemann's workload increased even further. In the meantime, it was clear that the Germans put in great effort to organize the transports to Westerbork, and from there on to the east. Whereas wide-scale exemptions from the requirement to report were initially granted, the number of exemptions dropped quickly in the autumn of 1942. Every day, hundreds of often desperate people gathered at offices of the Jewish Council with requests to be placed on a list that made it possible to get a temporary postponement from transport to Westerbork.

When in September it became known that most of the remaining *chalutzim* had been temporarily exempted from the requirement to report (so that they could contribute to the food supply), all sorts of 'more or less related' people

---

4 GFHA, Verklaringen Pinkhof. See also GFHA, Cat. no. 15, Verklaring Ilse Birnbaum, maart 1957, 1–4.
5 GFHA, Cat. no.13, Benjamin, Das Ringen, 12, 13.
6 Ibid.

began to show up at the pioneers' house. They asked Hannemann and his assistants for a place in the *hachshara* or, if that was not possible, addresses to hide. It was not unusual for 60 to 70 people to be present in *Beth Chaloets* on Sundays, including pioneers from elsewhere in the country.[7] This put great pressure on Hannemann and his helpers, while the stream of visitors also drew attention in the area and presented a security problem.

Other People Wanting to Go into Hiding

After the definitive closure of the *Werkdorp Wieringermeer* in the autumn of 1941, the JCB managed the general institutions of Loosdrecht, Gouda and Elden in the middle 1942. The situation in Loosdrecht was described earlier. In Gouda, the Catharinahoeve agricultural training center was set up in 1937, with a workforce of about about 25 pioneers. The Catharinahoeve was led by the Dutch-German married couple called Litten; in practice, Jansje Litten-Serlui, who came from Amsterdam, was in charge.

At the beginning of 1941, about 40 young pioneers between the ages of 15 and 17 years were housed in Huize Voorburg, a manor with a large orchard and meadows, located in the small village of Elden now part of Arnhem in the province of Gelderland. They had not yet completed their training and followed supplementary courses on fruit cultivation in the manor. The residents mainly originated from Loosdrecht and Gouda. The people at Elden were different from the other pioneers because they had family in Palestine, which meant that they possessed a Palestine certificate issued by the British government. Technically, this entitled them to settle in Palestine. However, this was impossible because of the war.

Five older pioneers were in charge in Elden. The most important leaders were Werner Ahlfeld (born in Nordhausen (Thüringen) in 1911) and the 41-year-old Dutch woman Linnie de Bruin. Ahlfeld, who came from a well-to-do family, had studied economics in Germany. After the *Kristallnacht*, he had been imprisoned for several weeks, after which he fled to the Netherlands at the beginning of 1939. Through Ru Cohen's Deventer Union, he then worked at farms in Dieren and Epse. His brother and mother lived in Palestine. Linnie de Bruin managed

---

7 Ibid., 14.

the household in Huize Voorburg. The pioneers at Elden, spent half their day following lessons, and working the rest of the time.⁸

In addition to organizing the general training courses, the JCB also managed two *kibbutzim* of the moderately religious *Mizrachi* movement in Beverwijk and Franeker in Friesland. In these kibbutz, the pioneers had a collective shelter from which they set off to work at the surrounding farms, similar to the situation in Loosdrecht. At the start of the occupation, the Germans closed the Beverwijk location because it was situated in the coastal area, where foreign Jews were not allowed. Of the approximately 100 members, some went to a new shelter in Laag-Keppel, near Enschede in the east of the country; others returned to their family. In the autumn of 1941, the *SiPo* raided the Franeker kibbutz, on the grounds that food was hoarded there. Two pioneers escaped. About 20 others were transported via Leeuwarden to Westerbork. Of these, six Dutch pioneers were released.⁹

In Enschede, there was another training center, the Haimers Esch, which belonged to the strictly orthodox *Agudah* movement. It had more than 50 members. This center and the shelter in Laag-Keppel continued to exist until the late spring of 1943. As organizations, they played no part in hiding *Hechalutz* people. Several orthodox leaders were convinced that the Jews had to accept their fate and that resistance was not allowed. In Laag-Keppel, where there were still about 40 pioneers at the beginning of 1943, the possibility was investigated of going into hiding collectively, as had been done in Loosdrecht. However, the leaders were not successful in finding sufficient places—although some individual members did hide.

Contrary to what might be expected on the basis of their convictions, a large number of the *Agudah* pioneers from Enschede did go into hiding. They acted on their own initiative and were helped by a group under the direction of the Dutch reformed minister Leen Overduin, who collaborated with the local Jewish Council. Almost half of them survived the war.¹⁰

Additionally, there were more than 200 pioneers from Deventer who took part in the individual *hachshara* organized by the Deventer Union. They worked mostly on farms in the east of the country. The pioneers were in contact with

---

8 GFHA, Cat. no. 1, Verklaring Werner Ahlfeld, oktober 1955, 1, 2. See also GFHA, Cat. no. 20, Verklaring Linnie de Bruin, december 1956, 1, 2.
9 Presser, *Ondergang, dl. 1*, 449. See also GFHA, Cat. no. 169, Verklaring Abraham Pach, januari 1957, 1, 2.
10 Pinkhof and Brasz, *De Jeugdalijah*, 12, 13. See also L.F. van Zuylen, *Palestinapioniers in Twente 1933–1945. Een vergeten hoofdstuk* (Enschede: Twente Akademie, 1995), 3–49. GFHA, Cat. no. 80, Verklaring Simon Gutmann, april 1957, 1–3.

each other at set times in the *Beth Chaloets* in Deventer. When the Jews were barred from public transport in June 1942, and when later they were no longer allowed to ride bicycles, the Union organized meetings in other places in the region, such as Zutphen. Generally, the organization of the Deventer group was much looser than that of the kibbutzim, as the training centers were so widely spread. The travel restrictions also limited the connection with Amsterdam.

For *Hechalutz* board member Hannemann and his assistants, those considered to be first in line for receiving help included the roughly 150 former *Werkdorp* pioneers, the training courses in Gouda and Elden and the individual pioneers of the Deventer Union. However, Hannemann regarded it as his task to help as many people as possible. He was supported in his work by his housemates Lore Sieskind and Thomas Drach, and by Norbert Klein. Drach and Klein were part Jewish. However, they had, similar to other so-called half-Jews, always been part of the Palestine pioneers. Drach had come to the Netherlands at the end of 1936; Klein emigrated in 1938, shortly after the *Kristallnacht*. Their status as half-Jews offered the advantage that they did not have to wear a star and could travel freely by train.

According to his own statement, Thomas/Gideon Drach, born in Munich in 1916, was first raised as a Christian and then became a socialist. After some fights with Nazi sympathizers in High school, he left for Lithuania in August 1933 to take part in the *hachshara*. Three years later he settled in the Netherlands and worked as a technician in different places. Drach, who had assumed the first name Gideon in Lithuania, had continued to stay in contact with the pioneer movement. For personal reasons, he had decided not to take the opportunity to sail to Palestine in the autumn of 1939 with the *Dora*. This was partly because he thought that he would be safe in the Netherlands.

After the German invasion, Drach went to work as an instrument maker with the NEDAP machine factory in Amsterdam. Although he was part Jewish, his identification card had initially been stamped 'J', meaning he had been registered as being Jewish at first. Later, at the beginning of 1943, he was able to obtain the status of half-Jew by going through the so-called 'Calmeyer procedure', which included an ethnographical investigation. Drach spent his time picking up ration cards, taking people into hiding and doing other courier work for Hannemann.[11]

Norbert Klein was born in Naumburg in Saksen-Anhalt in 1919. He had a Jewish father and a non-Jewish mother. He was therefore officially a half-Jew. After

---

11 GFHA, Cat. no. 70, Verklaring Gideon Drach, 1962/63, 1–5. With undated supplementary statements.

his arrival in the Netherlands at the end of 1930s, it had been possible for him to keep away from the activities of the Palestine pioneers. This was quite easy because he looked non-Jewish, with his blond hair and blue eyes. He had, however, joined the *chalutzim* and did the same resistance work as Gideon Drach.[12]

Lore Sieskind, who was 22 years old in 1942, came from Berlin and, just as many other pioneers, arrived in the Netherlands at the end of 1938. In Germany, she had been a member of *Hashomer Hatzair* and other left-wing Zionist organizations. In the Netherlands, she also belonged to the left wing of the pioneer movement, as evidenced by the economics course she followed given by the Marxist theoretician Sam de Wolff. She had extensive discussions with Shushu Simon, Hannemann and several others in the *Werkdorp* about the problems of socialism and Zionism.

During this time, Lore was friends with Hans Bandmann, who was one of the 59 detainees taken to Mauthausen in June 1941. Shortly after his arrest, Lore was summoned to the *SiPo*, who told her that her boyfriend had been shot while trying to escape. She then made a scene and said that she did not believe them, at which point they threw her out of the building.

Because Lore had seen how lists with the names and addresses of the pioneers to be rounded up were typed at the Jewish Council's office, she was convinced that cooperation with this Council would be dangerous. She therefore argued in favor of establishing contacts with Dutch socialists and communists.[13]

After the *Werkdorp* was closed, Lore Sieskind first worked as a housemaid for several wealthy Jewish families in Amsterdam; from July 1941 onwards, she was active as a child-care worker at a day nursery in the Plantage Middenlaan. The crèche was situated in the Jewish quarters, but was initially also attented by non-Jewish children. In the summer of 1942, the Germans turned the *Hollandse Schouwburg*, which was located opposite the crèche, in a deportation center for Jews to be transported to Westerbork. The Jewish Council then began using the crèche to house small children from the *Schouwburg*. Many children were smuggled out of the building, after which they were housed with non-Jewish families. From August 1942 onwards, Lore was in charge of the housekeeping of the *Beth Chaloets* in the Tolstraat, but she was also active in the same sort of resistance work as Drach and Klein.[14]

Harry Asscher, one of the new helpers, was not originally a member of the the *Werkdorp* pioneers. He was born in 1912 into a religious Zionist family living in

---

12 GFHA, Cat. no. 112, Gegevens Norbert Klein.
13 GFHA, Cat. no. 217, Verklaring Lore Sieskind, december 1955, 1–8.
14 Ibid., 2–4. See for the crechè, De Jong, *Koninkrijk dl. 6, eerste helft*, 352–354.

Amsterdam. After having finished the MULO (junior High school), he worked in an office for several years, but he later deciced to take part in the religious *hachshara* in Franeker. Due to a serious illness, Harry was declared unfit for the program. He then got a job at the agricultural training program for religious youths, for which he also went to Loosdrecht. There he met Channa de Leeuw, who was part of the staff; they married in October 1941.

Harry was responsible for finding employment for the religious Palestine pioneers from Laag-Keppel and other places. When their work permits were revoked at the end of 1942 and the beginning of 1943, Harry tried to find addresses for religious pioneers who wanted to go into hiding, in collaboration with the Westerweel group. Later, he was also involved in the care for other pioneers.[15]

A New Network of Helpers

At the end of 1942 and the beginning of 1943, Hannemann and his assistants formed a network of non-Jewish helpers that operated alongside the Westerweel group, doing their best to provide hiding places, forged papers and other items that were useful to the *Beth Chaloets* group. It remains somewhat of a mysterie how the first contacts were made; how the different people involved were connected, and what the precise connection was with the Westerweel group. However, it is certain that Gideon Drach, who had been in the Netherlands for a long time, played an important role here.

At the end of 1941 for example, Drach had become acquainted with nurse Antje, or Ans Roos, who was 26 years of age. She came from a Dutch Reformed working-class family with six children in Alkmaar, in the province of North-Holland. In the 1930s, her father, who took great interest in the minor Old Testament prophets, had begun to regard the Nazi movement as 'the Antichrist', primarily because of the anti-Semitism in their program. He had passed on this conviction to his children. While hitch-hiking through Germany, Antje had observed that anti-Semitism had taken firm root in the heart of many Germans.

The entire family was involved in various forms of resistance, almost from the start of the occupation. Antje's brother Kees, a policeman in Amsterdam, helped Jews escape during the *razzias (raids)* in Amsterdam's city center in February 1941. Antje Roos, who worked as a nurse in the Amsterdam *Wilhelmina Gasthuis* (hospital) later recalled in an interview that she 'was not so much

---

15 GFHA, Cat. no. 9, Verklaring Harry Asscher, ca. 1955, 1–3.

Fig. 11: Antje Roos (left) with an unknown girlfriend in the Beth Chaloets (1942)

sad—we were all sad—but furious, a sort of anger that made you do all sorts of things that you otherwise would not have dared to do' during the occupation.

In 1941, she hid an Italian communist who had fought in Spain in an apartment she shared with collagues. Later, a Jewish communist lawyer and his wife followed. The first raids in the summer and the autumn of 1942 made a deep impression on Antje. From the window of her apartment she could see the line of arrested Jews stretching over the Ceintuurbaan, 'walking very slowly with those soldiers on both sides with carbines on their shoulders, the children, the women, the men excruciatingly slowly.' Antje found it incomprehensible that other people simply walked past the line without paying any attention to it. She had never thought that such a thing could happen in the Netherlands.

Her name had quickly become known in Jewish circles through her activities to help people go into hiding, as a result of which more and more people asked her for help. One of these people was Gideon Drach's uncle, a refugee from Germany whom she helped find a hiding place. It was Gideon Drach who introduced Antje Roos to Hannemann and Sieskind. She was often in the *Beth Chaloets* and sometimes also slept there when her own flat was occupied by people in hiding.

Antje Roos's resistance work was a kind of family business: her parents' house—a small workers' home with limited space—functioned as a transfer center.

Antje's brothers took in people in need of a hiding place, and they approached distant relatives and acquaintances with the request to hide Jews as well. Their efforts were not always successful. Antje's middle brother would walk through a village and think, as Antje Roos recalled later: 'Where should I go to now?' The villagers would peer through the closed curtains and say to each other: 'Yes, there's that Roos brother walking about again, trying to to talk more people into taking in some Jews.'[16]

Hannemann and Drach had warned Antje Roos to be careful, and not to trust too many colleagues. Roos and the pioneers worked together until the beginning of April 1943. She was then arrested in connection to the raid carried out by the resistance on the Amsterdam Population register in which her brother Kees (the policeman) was involved. Antje Roos received a twelve-year prison sentence, which was later reduced to one year. Kees Roos was condemned to death in the trial against those who carried out the raid and executed.[17]

Another relief worker helping the pioneers group was the insurance expert Bob Jesse from Amsterdam. He had been active in the socialist youth movement AJC before the war. After this organization was closed at the beginning of the occupation, he had joined a group centered around the left-wing socialist newspaper *De Vonk*. They asked him to create false papers and it was through this activity that he came into contact with Kurt Hannemann, somewhere at the end of 1942 or the beginning of 1943. Jesse supplied him with false identification cards a few times, and also managed to hide Jews in the south of Limburg on several occasions. In addition, Jesse also worked for other resistance groups, such as the *Persoonsbewijzen Centrale* (Identification cards Centre), led by Gerrit van der Veen.[18]

Frans Gerritsen, a Jack-of-All-Trades in the Resistance

Making forgeries and undertaking other resistance activities took Bob Jesse so much time that he sent Lore Sieskind and Kurt Hannemann to obtain false documents from a study friend of his wife Dientje in the spring of 1943. This friend

---

16 GFHA, Cat. no. 188, Interview Yigal Benjamin-Antje Roos-Geerling, 19 oktober 1989. See also Verklaring Gideon Drach, 26 mei 1964. During her interview with Benjamin, Antje Roos became so emotional that she suffered a stroke and died the next day.
17 Ibid., Verklaring Antje Roos. See also De Jong, *Koninkrijk, dl. 6, tweede helft*, 712–734. The mother of Rudi Bloemgarten, who was one of the main Jewish perpetrators of the attack, was hiding in Antje's appartment.
18 De Jong, *Koninkrijk, dl. 6, eerste helft*, 102–107.

was the industrial designer Frans Gerritsen, who lived in Haarlem. A close bond soon formed between the two *chalutzim* and Frans and his wife Henny.

Frans Gerritsen, a 'handsome and energetic guy' was born in Apeldoorn into a politically conservative, liberal protestant milieu in 1915. After the war, he remembered that an aunt once gave him 10 cents to shout at a socialist parade in celebration of May Day: 'Get rid of those red devils'. The family moved to Zeist, where he attended primary school and High school. Frans then went to Amsterdam to study industrial design at the *Instituut voor Kunstnijverheidsonderwijs* (Institute for Arts and Crafts Education), which is now known as the *Rietveld Academie*. From the end of 1930s, he worked as a manager at a woodworking company in Amsterdam. Frans and Henny Gerritsen lived in Haarlem from 1941 onwards, where they had friends in left-wing and pacifist circles.[19]

Frans quickly joined the resistance through the influence of the *chalutzim* and Bob and Dientje Jesse. At Bob Jesse's request, he 'lost' his identification card and hid a number of Jews in his house: first a Polish communist and then Simon and Saapke Prins, Jewish acquaintances of Gerritsen, from August 1942. Gerritsen always kept his activities quiet, with an eye to safety. Even Bob Jesse was not aware of all that Gerritsen was doing.

Jesse also taught Frans how to make forgeries, an activity for which his experience as an industrial designer came in very useful. Their first forgery was made the same evening that Frans's wife Henny gave birth to twins. It was a document for the so-called 'Somme line', a broad strip of land along the river Somme in northern France, where the Germans implemented strict checkpoints to prevent refugees and escaped prisoners of war from getting any further into the country. Gerritsen and Jesse created a stamp, 'that had to be as worn as an existing stamp and that we left to dry on a glass plate for the night.' Jesse had to interrupt the work to call the doctor from a telephone booth for the delivery of Henny's babies.[20]

Later, Bob and Dientje Jesse came to Frans almost every evening, bringing stacks of identification cards to which changes needed to be made. For these cards Jesse worked with the Jewish photographer d'Oliveira, who used different lighting techniques to make people who wanted to go in hiding look as non-Jewish as possible in their photographs.[21] These photographs were then attached to

---

19 Paul Siegel, *Locomotieven*, 138. See also GFHA, Cat. no. 63, Interview Yigal Benjamin-Frans Gerritsen, 19 december 1989, 1–18.
20 Ibid., with supplementary commentary and statements by Gerritsen on previous versions of the interview. All versions autumn 1989.
21 This was almost certainly the Jewish architecture photographer Jaap d'Oliveira. He had a non-Jewish wife and could operate more freely.

'lost' or stolen identification cards. The documents were often stolen from unattended handbags in cafés or from clothing in dressing rooms of swimming pools and sports clubs. After the new photographs had been attached to the cards, the municipal stamps and registration numbers had to be redone or altered. Alteration was required if the numbers were registered as blocked.

Fig. 12: Frans and Henny Gerritsen with their daughter Mieke. The photograph was made some years after the war.

Gerritsen often worked on these assignments until the early hours. He then had to go out at half past five to be at work on time. He returned home at seven o'clock in the evening, where, after a quick meal, he started his forgery work again. In the spring of 1943, it became to much for his wife who decided to take the twins to live with her parents for some time. The four people who were in hiding with them were put up elswhere.

Frans Gerritsen used the time that his wife was not at home to build two other hiding places, in addition to the place he already had. One was hidden in the attic and could be accessed through a ceiling lamp. By pressing the switch in and tuning it, part of the ceiling hinged open, after which a wooden retractable staircase became visible that led to a hiding place under the guest room floor. The other hiding place was intended to store the archives, in which the data of about 300 people in hiding, for whom the Westerweel group was responsible, was kept in code.

Competent handyman Gerritsen made all sorts of other items to help the resistance, mostly out of wood. These included hollowed out wooden prison washing tags, in which messages could be transferred. He also hollowed out frames of painting, legs of chairs and wooden armrests of settees; these could be used to hide secret items.

Lore Sieskind and Kurt Hannemann went into hiding with Gerritsen in Haarlem at the end of spring 1943. They had to leave Amsterdam because the Germans were after them. The *Beth Chaloets* resistance center in the Tolstraat was in fact within the danger zone from the autumn of 1942. There was a raid by the *SiPo*, in which everybody present was arrested, except for Lore Sieskind, who was in bed with jaundice. Gideon Drach hid behind a curtain and escaped arrest.[22]

Afterwards, towards the end of 1942, the Germans employed an infiltrator. By sheer coincidence, however, she was not able to do much harm. At the beginning of March 1943, one of Hannemann's helpers saw a post in the resistance newspaper *Vrij Nederland* warning people against a certain 'Mrs. De Ruiter', who was known to infiltrate resistance organizations for the Germans. This woman, described in greater detail in chapter 10, had at that point already been active in helping the organization for about three months. She was introduced into the *Beth Chaloets* by Norbert Klein, who was not always careful enough.[23]

After her exposure, the pioneers left the Tolstraat as quickly as possible. Lore Sieskind went to a house on the Afrikanerplein in the Transvaalbuurt, which had been allocated to Jews. Hannemann and Drach lived near Lore. The illegal activities were moved to the attic of the JCB premises in the Hemonystraat. Two days after the move, a group of six *SiPo* agents raided Sieskind's new home and arrested her. She was interrogated for hours about the work carried out in the *Beth*

---

22 GFHA, Cat. no. 63, Interview Yigal Benjamin-Frans Gerritsen.
23 GFHA, Cat. no. 70, Verklaring Gideon Drach, 3.

Chaloets in the Tolstraat and her part in it. Because she had prepared a story, this went well, as did a heavy interrogation the next day.

Sieskind and Hannemann, who was overworked, decided to leave Amsterdam; through Bob Jesse they ended up with the Gerritsen family in Haarlem. Lore helped Henny Gerritsen with the household chores and later resumed her old job, working as a courier. Kurt Hannemann began to help Frans Gerritsen forging documents. After short period of rest, Hannemann resumed caring for people in hiding.

Later, the Gerritsens hid another two pioneers, Paula Kaufmann and Emil Windmüller, Kaufmann, who would play an important part in the resistance, was born in Poland in 1920, and had lived with her parents in Vienna from an early age. After the annexation of Austria in 1938, she fled to the Netherlands, where she stayed in the *Werkdorp* until the beginning of 1941. After leaving the Werkdorp, she hid at different places, including in a hollow in the woods. The *Hechalutz* then got her a job as a household assistant with the Gerritsen family; she travelled to France at the beginning of 1944.

Frans and Henny Gerritsen and the Palestine pioneers had an excellent relationship from the very beginning. Their collective left-wing background formed the mail unifying bond, just as it did for the Westerweel group. However, unlike Joop Westerweel and his employees, Gerritsen remained the type of a no-nonsense go-getter, who was also involved in resistance work, which mainly consisted of making forgeries for other illegal organizations.

* * *

The expansion of the Westerweel group's resistance work certainly did not happen spontaneously. Initially, the original core group (mainly from Rotterdam), was less than enthusiastic about the plans proposed by the *Hechalutz* to hide new groups of pioneers, such as the former *Werkdorp* pioneers and the pioneers from Gouda and Elden and the Deventer Union. Joop Westerweel and his crew felt that the expansion would be too great a burden for the rather limited group of helpers.

Reinforcement came mainly from the Palestine pioneers, who built up their own organization parallel to the Westerweel group; however, they also mobilized several new non-Jewish helpers. The most important of these was Frans Gerritsen, who initially worked mainly as a forger and offered a hiding place in his home, but was later also active in all sorts of other areas in the organization.

The cooperation that was achieved at the end of 1942—with the *Hechalutz* and later also with Gerritsen—led to a new structure within the Westerweel group. The original core group around Joop Westerweel continued to exist, but there were also the *Hechalutz* network and the Haarlem section around Frans

Gerritsen. The 'cement' between these three 'departments' was the group of Palestine pioneers who acted as intermediaries and made communication possible. This gradually resulted in a close mutual collaboration that might be referred to as 'the renewed Westerweel group'.

6 A Second Hiding Operation

Chapter 5 described which groups of pioneers were likely to be taken into hiding and who could help them, aside from the original members of the Westerweel group. This chapter describes what it was like to be in hiding or to be without a place to hide. To illustrate the hardships faced by individual Palestine pioneers during the war, we will zoom in on the stories of two of them: Hans Flörsheim and Paul Siegel.

The Failure of Elden

We start with the attempts of the *Hechalutz* to convince pioneers from the other courses to hide after 'Loosdrecht'. The difficulty experienced in convincing some of them to do this has already been shown, and the case of the *hachshara* in Elden bears this out.

According to a post-war statement by Werner Ahlfeld, the training course leader in Elden, he had several discussions with Shushu Simon and Kurt Hannemann in the spring of 1942. 'We agreed on the necessity of hiding the children of the Youth Aliyah'. The Loosdrecht children should be taken into hiding first. In Ahlfeld's opinion, Elden was safe, for the time being, 'because the inmates were regarded as farm workers'. In the meantime, the leaders there, would look for people in the neighbouring area who were willing to hide Jews.[1]

This version is not very credible. Simon and Hannemann did not start to busy themselves with the other training courses until after the Loosdrecht pioneers were safely in hiding. After the war, co-leader Linnie de Bruin said that searching for addresses in Elden was not done systematically. Her statement was supported by Hans Seeman. He was born in Berlin in 1923 and taken to Eindhoven at the beginning 1939 on a '*Kindertransport*' (children's transport). From Eindhoven, he went to Elden via Loosdrecht. After the deportations had started in July 1942, he found a hiding place on his own. However, the leaders absolutely did not want individual pioneers to go into hiding, in order to avoid possible reprisals. Ahlfeld may have believed the German promise that 'Elden' would be allowed to continue their work, so Seeman abanboned his plans.

Ilse Birnbaum, who was 21 years old and also came from Berlin, was also not convinced that Ahlfeld and De Bruin had an 'overall plan' to let the pioneers

---

[1] GFHA, Cat. no.1, Verklaring Werner Ahlfeld, 2.

hide. Ilse, who was part-Jewish and could therefore act more freely, looked for addresses where her boyfriend Ernst Cosmann (an older pioneer who assisted the leaders at Elden), Linnie de Bruin and her daughter could hide. Out of solidarity with the others, Cosmann and De Bruin refused to use these addresses. Still, four or five Elden pioneers did go into hiding in the summer of 1942.

According to Linnie de Bruin, the Deventer Union intended to send a warning if the Germans took action. In the event of such a warning, they could still go into hiding collectively. However, the Germans, who had learned their lesson from their Loosdrecht experience, acted without any warning against the Jewish organizations. In the early morning of 3 October 1942, a Dutch MP platoon, led by several SiPo agents, surrounded the Elden manor and transported the approximately 40 inhabitants to Westerbork.[2]

Fig. 13: The hachshara pioneers in Elden in the summer of 1942. Leader Werner Ahfeld is standing on the far right.

As Seeman suggests, Ahlfeld probably had to much faith in the German promise that the people at Elden would be allowed to carry on working because of the contribution to the food supply. The combination of this promise and the

---

2 GFHA, Cat. no. 20, Verklaring Linnie de Bruin, 1, 2. See also Cat. no. 213, Verklaring Hans Seeman, maart 1957, 1–3. Cat. no. 15, Verklaring Ilse Birnbaum, maart 1957, 1, 4. Cat. no. 15 Interview Yigal Benjamin-Ilse Birnbaum, 8 april 1987, 1–4.

Palestine certificates, which were highly valued in Jewish circles during the war, had lulled the leaders of the Elden group into a false sense of security, and had caused them to think – erroneously – that going into into hiding was riskier than waiting.

At Westerbork, they had a stroke of luck: the father of one of the group members, Manfred Samson, happened to be one of the camp inmates who had been there longest. This made him an influential person with the German camp leaders. Father Samson ensured that the deportation of the people from Elden was 'deferred'.[3] At the end of 1942, they were allowed to form a *hachshara* group that could work on agricultural projects outside the camp. Later, they were involved in all sorts of jobs inside and outside Westerbork. Other pioneers who entered Westerbork continued to join them and the group started to form a resistance core, which assisted in activities such as escapes. Chapter 9 contains more about the *hachshara* group in Westerbork.

The Pioneers in Gouda

A second group of pioneers who were considered candidates to go into hiding stayed at the Catharinahoeve in Gouda. The Hoeve was founded in 1910 as the '*Stichting voor Joodse tuinbouw, veeteelt en zuivelbereiding*' (Foundation for Jewish horticulture, livestock farming and dairy preparation). It was not until 1937 that this became a reality, when a small agricultural company of two hectares (about five acres) was set up with the help of a private donation of 80,000 guilders. According to the British-Jewish writer Jakov Lind, the house was surrounded by wide moats and was only accessible over a drawbridge like a 'ship stranded between the beans and tomatoes'. After May 1940, Lind stayed in Gouda for quite some time, though with great reluctance. In his opinion there was too much religious influence.[4]

After the *Kristallnacht*, the farm was home to the children of the *Jeugdalijah* (Youth Alyah*)*. Jansje (later Shoshanna) and Manfred Litten were in charge. Jansje was mainly concerned with taking care of the pioneers, while Manfred, a philosopher with a doctorate, taught Jewish subjects and Ivrit. There were also two non-Jewish 'assistants', who taught vegetable and fruit cultivation. Dirk van Schaik, a foreman at the local candle factory and passionate amateur

---

[3] Ibid., Verklaring De Bruin. See also Cat. no. 25, ongedateerde Verklaring Betty Baars, 1–4.
[4] Jakov Lind, *Stap voor stap* (Amsterdam: De Bezige Bij, 1970), 69. Lind (pseudonym of Heinz Landwirth) who was born in Vienna in 1927. He went to the Netherlands at the end of 1938, together with his sister.

gardener, worked there in his spare time. He would later play an important role as one of the group helpers, together with assistant instructor J. Middelburg.

Most of the twenty-odd pioneers, who stayed at the *Hoeve* in varying groups, had left Germany and Austria at the end of the 1930s and were about 15 years old when they arrived in the Netherlands. There were about seven girls. Similar to the situation in Loosdrecht, the living circumstances at the 'youth farm' were very simple. There was limited space and they slept in bunk beds in a dormitory. In addition to the foreign refugees, there were also several slightly older Dutch pioneers. One of them was Berrie Asscher from Amsterdam, who was 19 years old when he arrived in Gouda in September 1939.

Berrie came from a religious Zionist family. After elementary school, he followed a teacher training program at the Jewish seminary for several years. However, in part through force of the circumstances, he chose the *hachshara*, just as one of his older brothers had done. As a city kid who had mainly been studying books, he had to get used to the very different rural life. This involved normal agricultural work such as weeding and using a hoe. Two-thirds of the farm consisted of greenhouses and open boxes in which mainly vegetables were grown. After the harvest had been brought in and the vegetables had been packed into boxes, there were three transports a week by flat-bottomed boat to the auction in Reeuwijk near Gouda. Transporting the vegetables was a job for the older boys, such as Berrie.[5]

Shoshanna Litten had warned all pioneers: 'If you are arrested by the Germans, your last hour will soon have sounded'. To keep them from being sent to Westerbork, she had contacted Hannemann early on. She had also tried all possible ways of collecting enough money to finance hiding the pioneers.

The difficulties in Gouda started at the end of 1942. Five boys had to report for a Jewish work camp in the Netherlands. Because it was considered unwise to make all of the pioneers go into hiding just because of these five boys, doctors helped them to get admitted to the local hospital as dysentery patients. The boys stayed there for more than a month until the danger had passed.

A number of fanatical members of the Gouda NSB posed another threat. They were continually monitoring that everything went according to the anti-Semitic ordinances. Any violation of the rules at the Catharinahoeve, such as pioneers being outside after the eight o'clock curfew, were immediately reported to the Gouda police. When someone was arrested for misbehavior, Shoshanna had

---

5 Berrie Asscher, *Van Mokum naar Jeruzalem (1924–1944)* (Beersheba: self-published, 1996), 1–81.

to act quickly to get the pioneer released before he or she was sent to Westerbork as a 'punishment case'.

In this way, she succeeded in keeping all the pioneers out of the hands of the Germans until 22 April 1943. On this date, the inhabitants of the Catharinahoeve were ordered to report to the newly-opened concentration camp at Vught, in the southern province of Brabant. Under Shoshanna's leadership, seventeen pioneers packed their belongings and left for the train station. They bought tickets to Vught, and split up in groups. One after the other, the groups disembarked the train several stops after Gouda to go to different hiding places. Shoshanna had found the addresses with the help of Dirk van Schaik, who had become an increasingly important helper of the Gouda pioneers.[6]

Shoshanna Litten, who had been separated from her husband and later also from her son during her time in hiding, started to care for the young pioneers. One of them was Alfred Dubowski. He was born in German Koningsberg (today Konin in Poland) in 1926 and came to the Netherlands in November 1938 as a 13-year-old boy. He went to Loosdrecht in October 1940, but was unhappy there. He found the *Paviljoen* too big and too busy. Gouda, which was smaller, was much more to his liking.

Shortly before the other pioneers started their so-called journey to Vught, Alfred went into hiding with a Catholic worker's family in April 1943. He had been there for several weeks when Shoshanna suggested that he should go to France. He agreed, and she gave him a false identification card and money that came from the *Hechalutz* in Amsterdam. On 21 May, he travelled from Gouda to the Rotterdam Central Station, where he met three other pioneers, two of whom had also worked in Gouda for some time. The fourth was Emil Glücker, who had escaped several months before when he and a group of pioneers from Westerbork had to help clean up the Jewish mental institution *Het Apeldoornse Bos*, which had been closed by the Germans.

These four men were accompanied from the Central Station by 'Theo', a Belgian smuggler, about whom there is more in the next chapter. Theo was dressed in the uniform of the German civil and military engineering group, *Organisation Todt* (*OT*; more on *OT* also in later chapters), he took Dubowski and his comrades to Belgium in a detainee compartment of a German army wagon. They were posing as Dutch 'workers' who were being taken back to their work in France under his supervision. Theo disappeared in Mechelen, but left them with instructions

---

6 Benjamin, Manuscript Das Ringen, 15, 16. Several pioneers went to their parents shortly before. See also M.J. van Dam, H.C. van Iterzon and H.J. Maarsingh, *Gouda in de Tweede Wereldoorlog* (Delft: Eburon, 1995), 169–184.

on how to cross the Belgian border at Charleroi to get to France. The plan worked; Alfred reached Paris on 23 May 1943.[7]

Berrie Asscher, who returned to Amsterdam in November 1942, had also fled with the help of Dirk van Schaik. Dirk picked him up in Amsterdam at the end of May 1943 and took him to Reeuwijk, where Shoshanna Litten was hiding with her son in an attic. After two days, Berrie went to Rotterdam accompanied by another pioneer. From there he went to Belgium, after some difficulties at the border.[8]

Of the group of 21 Gouda pioneers, 15 survived the war. Shoshanna and Manfred Litten were arrested and murdered in Poland. Their son survived.

Fig. 14: A small group of the Gouda pioneers on one of the boats

There is another report of pioneer life in Gouda (or more accurately, of the life of someone in hiding in Gouda and in Amsterdam after that), in which the problems and the tensions associated with a life in hiding clearly come to the fore. This report was written by Hans Flörsheim, who was born in the small village of Rotenburg near Kassel (Germany) in 1923. In 1937, his reasonably well-off parents sent him to family in Amsterdam because of the increasingly worsening

---

7 GFHA, Cat. no. 8, Verklaring Alfred Dubowski, april 1956, 1–3.
8 Asscher, *Van Mokum*, 115–121.

climate for Jews in Germany. Hans obtained his HBS diploma (*Hogere Burgerschool*, High school) at the Jewish Maimonides Lyceum in Amsterdam. Not entirely sure what to do after school, he went to the *Werkdorp* with two friends in the autumn of 1940. Here, he belonged to the group that was allowed to help with the harvest until September 1941. He then followed his *hachshara* training in Gouda.

He stayed at the Catharinahoeve for some time, but then looked for a room and work elsewhere, although he often went back to eat meals. Flörsheim had absolutely no illusions about the German intentions. His father, who had been an active member of the socialist party, had been imprisoned for some time after the *Kristallnacht* and was arrested again later. In December 1941, Hans received a message that his father had died from a heart condition, which was a euphemism often used for a violent death. Shortly afterwards, his mother together with other Jews was taken from Leipzig, where she lived, to the ghetto of Riga. He never heard from her again.

Hans Flörsheim, who had found a job in the Jewish old peoples' home of Gouda, vowed never to go on transport voluntarily. 'I do not know when I had these thoughts for the first time, but one thought kept going through my mind: Dear God, please do not let me be sent away as well!'[9]

At the same moment, he realized that he was trapped in Gouda by this time, it was February 1943. Every day he lived in fear of being arrested in the street, despite his non-distinct Jewish appearance, because of his Jewish badge. His other great fear was that the old people's home would be cleared out. Flörsheim considered fleeing to Switzerland, but he did not have the money. By coincidence, he went to the *Beth Chaloets* in the Tolstraat during a weekend off, because a colleague had asked him to take a letter to someone there. He had never heard of the place before. The person to whom he had to give the letter told him that Hannemann, who Hans Flörsheim vaguely knew from the *Werkdorp*, might be able to help him to flee.

He spoke to Hannemann, but he 'did not say much, I still remember today, and his attitude discouraged me a great deal. I had the impression that he was only half listening. But that was just part of who he was, just like his thin, pale face was'. Hannemann told Hans that he 'would be hearing from him —and he did. At the beginning of March 1943, Norbert Klein, Hannemann's assistant, brought him a forged passport in the name of Hendrik Westerman, bricklayer.

---

[9] Hans Flörsheim, *Über die Pyrenäen in die Freiheit* (Konstanz: Hartung-Gorre Verlag, 2008), 15–52.

He also gave him a forged permit to travel by train. Flörsheim had to use this 'the moment anything happened'.

That moment came on the evening of Friday 9 April, when the director warned that the Germans were going to clear out the old people's home in half an hour and transport the inhabitants. Hans Flörsheim had not prepared himself for this situation and ran away in sheer panic. He first went to his room to organize some things and to think about the situation. From his window, however, he saw that the Dutch police were ringing the bell at the house of his Jewish neighbors to take them away. 'At this moment I was completely at a loss what to do. I had only one thought: Get out of here, to freedom!' He walked outside through the backyarde, tried to cover his star and, went to the house of a non-Jewish former colleague. However, the father of this colleague did not want to hide him. Flörsheim then suddenly recalled the address of a black marketer with whom he done some business; he was able to spend the night in his shop.[10]

It was not until several days later that Hans dared to travel to Amsterdam without a star. Here, he once again met with Hannemann, who gave him an address to hide in the guesthouse of the German, non-Jewish Mrs. Heiman. During the first fourteen days Flörsheim had to force himself to walk for two hours every day. 'It seemed as if everybody was looking at the spot where the star had been.' Later, as a sort of spectator, he closely followed the last German measures to make Amsterdam 'Judenrein' (free of Jews):

> On that day, I was awoken at 4 o'clock in the morning by the screaming sirens of police vans. Shortly afterwards, the actual raid started and agents entered every home to check papers or to search for forbidden items. My roommate and I were trembling, while our landlady worked wonders with her German pass.

In the meantime, the core members of the *Beth Chaloets* had gone into hiding elsewhere (Hannemann and Sieskind) or had been arrested (Drach and Klein), Hans helped Kurt Reilinger and Lore Durlacher, who had succeeded Lore Sieskind, with resistance work. Durlacher, who was 23 years old, came from a small place near Mannheim, where her parents owned a shoe shop. She arrived in the Netherlands in August 1939, and went to work in the *Werkdorp Wieringermeer*. After its closure, Lore went to *Het Apeldoornse Bos* as a student nurse. Shortly before this psychiatric hospital was cleared out at the beginning of January 1943, she had gone into hiding in Hilversum and later in Zaandijk. As a result of her clear-headed actions during several police raids at different hiding ad-

---

10 Flörsheim, *Über die Pyrenäen*, 55–60.

dresses, Lore, who looked like a 'naive young girl' with her hair dyed blonde, had gained the reputation that nothing could affect her.[11]

The three of them worked well together until the beginning of September 1943, when the *SiPo* raided Mrs. Heiman's guesthouse again. Her German papers did not help this time, and she was arrested along with everyone living there.

Hans Flörsheim, however, did not give up. When he was allowed to grab his coat, he quickly walked to the second-floor balcony. He saw immediately that jumping down to the street was too dangerous. 'Still, I was on the right path! I started to think and act almost simultaneously. I jumped on to the wooden railing, stepped around the wooden wall that separated our balcony from that of the neighbors, and was out of sight from our home.' In this manner, he slid over a few balconies and reached the street through an apartment.

Later, 'ice queen' Lore Durlacher picked up illegal papers from his former hiding place and provided him with a different identity card, with which he could go into hiding again. A month later, at the beginning of October 1943, Menachem Pinkhof helped Hans Flörsheim and a friend across the Belgian border near Breda. The two of them reached Antwerp by bus and tram.[12]

The Isolation of the 'Deventer' pioneers

This chapter devoted special attention to what happened to Hans Flörsheim, since he represents a different type of pioneer in hiding than the members of the Loosdrecht and Gouda group. He was slightly older and operated independently, which caused a great deal of tension. Under no circumstances did he ever want to go to Westerbork, but in Gouda he appears to have been paralyzed by fear. This prevented him from preparing for the closure of the rest home. He gradually shook off this fear in Amsterdam and even lent a helping hand to other Jewish pioneers in their resistance efforts. His escape showed his courage and level headedness.

It is curious that Flörsheim did not become aware of the illegal activities in the *Beth Chaloets* until late in the day and then only by accident. After the pioneers of Loosdrecht went into hiding, these activities seemed connected primarily to the *Werkdorp* pioneers who stayed in Amsterdam and the surrounding area.

This is also evident from the story of Paul Siegel. He was born into a middle-class Cologne family in 1924. His father was a poultry dealer and did lots of busi-

---

11 Siegel, *Locomotieven*, 138.
12 Flörsheim, *Über die Pyrenäen*, 65–74.

ness with the Netherlands. Shortly after Hitler had been elected Chancellor of Germany, Siegel's father decided to leave with his family. They lived in Oosterbeek, and later in Arnhem, where Paul attended elementary and High school. Influenced by friends, he became a Zionist, even though his parents were not. At the end of the 1930s, they declined an offer to emigration to Palestine, and chose to wait for visas for Canada, but these visas never came. After the German attack on the Netherlands, Paul decided to leave school and he applied for the individual *hachshara* at the Deventer Union. In August 1940, after initially having been rejected, he started his training with farmer Brekveld in the hamlet of Oeken near Brummen, about 20 kilometers from Arnhem.[13]

Paul Siegel, who had the appearance of a 'Yemenite Jew', felt very much at ease with the simple farmer's family who did not have electricity or gas. He often heard day laborers who helped with the harvest, talk in amazement about 'the Jew who works just as hard as we do'. He explained to them that Jews often also did heavy-duty work and that not all of them were cattle dealers or merchants, which was what many people in rural communities thought.

Dozens of young Jews worked as individual pioneers with farmers in the region along the IJssel river. During the winter, when there was less work, they would meet in the community center in Deventer once a week in the afternoon to attend lectures about agriculture and to talk with each other. Later, there were also meetings in the Union's *Hechalutz* house in Zutphen. Paul still regularly visited his parents in Arnhem and was therefore well aware of the anti-Semitic measures which were forcing Jews into an increasing state of isolation.

In the remote village of Oeken, where he had been working for two years without any problems, Paul was surprised one day in the summer of 1942 by a summons to report for a Jewish work camp. He decided not to go, but to ask good friends of his parents for a hiding place. However, they refused.[14] The Deventer Union helped him get a deferment from reporting until the end of the winter, but Paul did not use this. Instead, at the beginning of October, he went into hiding with three friends in a shack on a nearby estate. The owner, a baron, had given them permission to do so. In Paul's words, it was 'a Karl May-like undertaking', which went well in the first few weeks, because of good weather.

Siegel and the other pioneers decided on their adventure partly because none of them knew what to do. There was hardly any contact with the *Hechalutz* that worked from Amsterdam and 'no sign from Deventer indicating the exis-

---

13 Siegel, *Locomotieven*, 7–45.
14 They did hide Paul's sister later.

Fig. 15: Eight pioneers of the Deventer Vereniging in Assen in February 1942. Several pioneers worked there temporarily in a paper factory. Max Windmüller can be seen on the far left, Kurt Hanneman is the fourth from the left, and Schraga Engel is far right.

tence of a resistance organization'. However, their stay on the estate did not last long. The weather became colder, one pioneer left and they were warned about a rumor going round the village about Jews who were living in the woods. At the beginning of November, the baron said that they had to leave immediately because he was afraid of German reprisals if they were discovered.[15]

Paul Siegel walked to Deventer to consult with Ru Cohen, the Union's leader, for whom he had great admiration, but this did not help him much. Cohen, Siegel noted, 'believed the explanation of the Germans that the Jews would be housed in Poland according to a new plan. He regarded what he had heard about the Jews that had been sent to Mauthausen as a trick to deter us'. Everybody who had been transported there was still alive. Cohen did not believe in going into hiding, although everyone who did could count on his support. In this regard, he also told Paul about the *Hechalutz* resistance organization. However, Paul decided not to trouble the organization, which was still in the processs of being set up.

---

15 Siegel, *Locomotieven*, 62–77.

Paul Siegel now went in search of new addresses, ending up at the farm of a strict Christian family in a hamlet near Brummen, not far from Oeken. There were already twelve Jews in hiding there, including several pioneers. Paul realized immediately that this plan was also doomed to fail, because security was completely lacking. There were random passers-by, black-marketers and people in hiding on the farm. Several days after his arrival, the Dutch police raided the farm in mid-November 1942. A policeman from Brummen who knew Paul gave him the opportunity to escape. However, discouraged by the series of setbacks he did not take it. At the end of November 1942, he arrived in Camp Westerbork via the Arnhem prison.[16]

Paul Siegel's lack of contacts and subsequent search for a place to hide were not an exception. Other pioneers who fell under the Deventer Union's individual *hachshara*, were also unaware of the existence of the *Hechalutz* resistance organization. Ilse Birnbaum, who was discussed in relation to the Elden *hachshara*, was involved in the *Hechalutz*, where she cared for the Deventer pioneers in the spring of 1943. At that time, she worked in Twello near Deventer for a vegetable grower and began to care for four married couples of pioneers who were in hiding nearby. One married couple, Leni and Werner Rose, 'lived under the most pitiful circumstances and had no contact with the *chaverim* until the mid-1943s. They had no food ration cards, no money, no clothes and Leni was pregnant. Their hosts were terribly poor; he was a day laborer for a big farmer, who paid them in food for their work', Ilse remembered.[17]

Individual Hiding and Restored Contact with the *Hechalutz*

Gerd Schönebaum came from a very modest social background. He was born in Dortmund in 1919, and moved to the Netherlands at the beginning of 1939. Here, he worked as one of the Deventer Union's individual *hachshara* pioneers for a farmer in Almelo near Enschede. When he was summoned to report for a Jewish work camp in June 1942, he decided to go into hiding. Gerd, who looked Jewish and did not speak Dutch very well, then lost contact with both Deventer and the *Hechalutz* in Amsterdam. The result was that 'each of the individual hachshara members had to look after themselves and had to act on their own. We were not prepared for the call to go to a work camp'.[18]

---

16 Siegel, *ibid.*, 78–85.
17 GFHA, Cat. no. 15, Verklaring Ilse Birnbaum, 2, 3.
18 GFHA, Cat. no. 210, Verklaring Gerd Schönebaum en Toos de Leeuw, februari 1957, 1–4.

Gerd temporarily hid at the home of his Jewish girlfriend Toos de Leeuw in Almelo. A short time later, an unknown person appeared and offered help. This stranger was Derk Senechal, an anarchist, conscientious objector and leader of a local group of free socialists. According to Schönebaum, who considered him his friend, Senechal helped 'out of purely humanitarian motives'. Senechal 'was very interested in all possible subjects, but was poorly educated. He was strongly anti-Zionist and helped the Jews, but denied that there was a racial problem'.

Senechal, who was born in Groningen in 1913, was a baker and was married with three children. His wife helped him with his resistance work. In their political views and working methods, the Senechals showed marked similarities with Joop and Wil Westerweel, although there is no evidence suggesting that they knew each other. In August 1942, Senechal took Gerd Schönebaum to a socialist worker's family in Almelo. Another pioneer, Max Rothschild, was already hiding there. They stayed with the family for more than one and a half years.

In the autumn of 1943 they were visited by Lore Durlacher, who offered them the chance to flee to Spain. However, Gerd and Max did not accept the offer, 'because we believed that we were safer while in hiding'. This decision was made in part because Gerd believed that in his case 'the Hachshara organization activity (started) much too late'.[19]

Because of conflicts arising from tensions caused by a life in hiding, Gerd found a new place to live with a socialist laywer, at the beginning 1944. He was liberated at this address more than a year later.

Gerd's girlfriend Toos de Leeuw went into hiding in November 1942, also with help of Derk Senechal's group. She had to move to new hiding places eleven times, for different reasons.[20] Toos and her parents, who were also helped by Senechal, survived the war.

The statement of Greet Stork, born to Jewish-Dutch parents in Hamburg in 1917, supports the idea that the Amsterdam *Hechalutz* seemed to initiate little or no contract with part of the Deventer pioneers until the mid-1943s. Greet and her boyfriend Heinz Herz worked for farmers near Dieren between Arnhem and Zutphen. At the end of 1942, their plan was to stow away on a coaster bound for Sweden. Herz knew a sailor who was prepared to hide them. Unfortunately, despite all their preparations, nothing came of the plan.

At the beginning of 1943, the Deventer Union sent Greet an exemption from reporting at Westerbork. This allowed her to continue working on the farm where she was staying. At this time, she was helped by members ot the resistance in

---

19 Ibid., 3.
20 Ibid., 3, 4.

Dieren, with whom she had also hidden for some time. In June 1943, contact with the *Hechalutz* was restored and Greet Stork left for Amsterdam to wait for a definite hiding place. In the autumn, Lore Durlacher took her to Alkmaar, where she lived for some time at an address arranged by Antje Roos's family. After the police raided this address, Greet fled to the home of Antje Roos's parents. They had continued their resistance work, despite their son's death and Antje's imprisonment. Harry Asscher then took Greet to Witmarsum in Friesland, probably at the beginning of 1944, where she hid with a doctor and his wife. Joop and Wil Westerweel's daughter Ruth was already hiding here, together with two other Jews.

After a stay of several months in hospital, Harry Asscher again took Greet Stork to a minister in Bolsward. He himself had also been in hiding there previously. She stayed here until the liberation. Her boyfriend Heinz Herz, whose story appears later in this book, was murdered in an extermination camp.[21]

The Failures and Sucesses of Going into Hiding

The situation described above paints a mixed picture of the resistance activities of the *Hechalutz* board after 'Loosdrecht'. The 'hiding operation' in Loosdrecht turned out to be successful due to a combination of factors, which will be the focus of this section. These factors also determined the success or failure in hiding the pioneers in other training courses. The *Paviljoen's* leaders were sympathetic towards pioneers going into hiding, and proactively started to make sure that necessities, such as money, false papers and addresses to hide were provided. When they found that they were not able to do this on their own, they called on several non-Jewish helpers who, with Joop Westerweel as their great inspiration, supplied them with what they were not able to arrange themselves.

The majority of the Gouda pioneers went into hiding following the same pattern. The leaders actively contributed by collecting money and papers according to the 'Loosdrecht' model. They were successful in mobilizing a few non-Jewish helpers who provided addresses to hide. Dirk van Schaik, who led this initiative, even accompanied several former Gouda pioneers, who were in hiding elsewhere to the border. The modest Gouda network was assisted by members of the Westerweel group, as well as by newly recruited helpers such as Frans Gerritsen, who provided papers.

On the other hand, there was the failure of Elden, where the leaders supported the idea of going into hiding in principle, but made no serious effort to make

---

21 GFHA, Cat. no. 272, Verklaring Greet Stork, 1955, 1–3.

the necessary preparations. They probably placed too much trust in German promises and Palestine certificates. Only four pioneers from Elden did go into hiding, even though the activities of Ilse Birnbaum, who found several addresses to hide, showed that there was a potential network of helpers.

Moreover, the lack of action from the leaders of Elden led to the large number of arrests when the manor was raided by the Germans, who did not announce their action to the Deventer Union beforehand, as had previously been agreed. This meant that the warning system did not work. However it remains unclear what the significance of this warning would have been, considering the fact that there were hardly any preparations in place for going into hiding.

Despite the fact that the *Hechalutz* board had no contact with the religious training centers about going into hiding, several pioneers from these centers received help from them. For example, Ab Pach, born in Amsterdam in 1923, escaped the German raid on the religious *kibbutz* in Franeker in the autumn of 1941. After having legally worked for a Jewish butcher in Rotterdam, he went into hiding with a false identification card. He stayed at nine different addresses, which he go for some time from the *Hechalutz*. He had to leave his last address in Hillegom near Haarlem because he was betrayed, after which he went to work as a non-Jewish farmhand in Brabant in May 1943. He lost contact with the *Hechalutz* as a result. At the beginning of 1944, he unexpectedly received a visit from Harry Asscher, who had gotten his address in Hillegom. He offered Pach the chance to flee to France. Ab Pach accepted this offer, also because the farmer that he worked for often spoke negatively about Jews. In March 1944, he travelled to Paris with false papers given him by Menachem Pinkhof in Rotterdam.[22]

Several religious pioneers from Laag-Keppel were also helped by the *Hechalutz*, although it is not clear how many.

The Hechalutz also helped *Werkdorp* pioneers, who stayed mainly in Amsterdam and surroundings. After the *Werkdorp* was closed, they remained in contact with the umbrella organization, which provided them with false papers and addresses to hide. Antje Roos in Amsterdam and Frans Gerritsen in Haarlem set up a network of non-Jewish helpers though they did not exclusively focus on the Palestine pioneers. In this manner several dozen *chalutzim* were saved.

It should be taken in consideration that the *Werkdorp* had been largely closed by March 1941, almost one and a half years before the *Hechalutz* began to actively help people who wanted to go into hiding. Furthermore, the Germans sent groups totaling more than one hundred *Werkdorp* pioneers to extermination

---

22 GFHA, Cat. no. 169, Verklaring Ab Pach, 12.

camps in June 1941 and July 1942. Many pioneers were arrested later during a *razzia of* an agricultural project near Amsterdam at the end of May 1943.

The precise number of Deventer Union and *Hechalutz* pioneers that were able to go into hiding is also unknown. It probably amounts to several dozens, possibly almost a hundred people. This would likely have been more if the leaders of the Union, and especially the founder, the very influential Ru Cohen, had chosen to cooperate. However, he advised against going into hiding for a long time, as Paul Siegel described. With regard to the Deventer Union, no preparations were made for building a network of non-Jewish helpers.

Hans Mogendorff, who was born in Düsseldorf in 1922 confirmed this passive attitude. He settled with family in Eindhoven in 1937 and, as a passionate Zionist, went on individual *hachshara* in Ru Cohen's furniture shop in Deventer in the autumn of 1940. After a year, Hans went to work for a farmer in Usselo, near Enschede, where he was summoned to report for a Jewish work camp in the Netherlands in August 1942. He did not like the idea of the work camp and obtained a false identification card through his girlfriend who had connections with a Nijmegen resistance group. Before going into hiding Hans asked Ru Cohen for advice, who told him not to do it. In 'afrank and heated telephone call,' he informed Cohen that he was going forward with his plan.[23]

Cohen and several prominent Dutch Zionists, such as Sam de Wolff and Abel Herzberg, had no idea of the ruthless nature of the German anti-Semitism. Ru Cohen believed the promises of subordinate German authorities that his pioneers were safe, as long as they did their work on the farm and did not go into hiding. He could in fact not accept the downfall of his *hachshara* work, which had been built on so much effort and idealism. In the spring of 1943, it became clear that the Germans did not keep their promises, and Ru Cohen finally went into hiding with his wife and two daughters. By then, he was a broken man. The family reported voluntarily in the autumn of 1943, when the *SiPo* arrested their third and oldest daughter, Rachel, in Rotterdam. Ru and his wife died in Bergen-Belsen. Their two youngest daughters survived the war.

There were also core groups of helpers, like Derk Senechal and kindred spirits in Almelo and the resistance in Dieren, in the working area of the Deventer Union. However, at the end of 1942 and at the beginning of 1943, the number of Dutch people who were willing to help Jews go into hiding was generally limited. Here, it is important to remembered that the Landelijke Organisatie voor Hulp aan onderduikers (LO) (National Organization for Help to People in Hiding)

---

23 N. Mageen (Hans Mogendorff), *Van zonsondergang tot dageraad* (Beersheba: self-published, 2002), 8–19.

was still being set up. Ilse Birnbaum, who worked hard to find possible hiding spots for the Elden group and later Deventer pioneers during this period, reported in 1957 that it 'was very difficult to find hiding places. I often had the impression that the Dutch could have helped more, but they were sometimes not interested'.[24] This was worded very carefully.

The Dutch resistance in the east of the country lacked the inspiration of someone like Joop Westerweel who, together with the resolute leadership of the Union, could almost certainly have accomplished more.

An important disadvantage was that the pioneers were spread over a large geographical area, which made both internal communication and communication with the *Hechalutz* difficult and hindered decisive action. An additional problem was of course the fact that all of the work involved in hiding people fell on the shoulders of a small group of people, who all became overburdened after a while.

---

[24] GFHA, Cat. no. 15, Verklaring Ilse Birnbaum, 2. Birnbaum stated that when help was offered, it was mostly offered by people from religious circles.

7 The Search for Safety

From the very beginning, the Westerweel group had intended to take the pioneers who were in hiding to the safety of Switzerland. It was clear that it would be very difficult in the Netherlands to find sufficient addresses to hide, among other reasons because the pioneers were conspicuous, due to their German accent and/or Jewish appearance. However, an attempt that was made to take eight pioneers to Switzerland with the aid of people smugglers, in October 1942, ended in failure—so that other routes were required.

After the successful 'hiding operation' of Loosdrecht, youth leader Shushu Simon sent a coded letter to Nathan Schwalb, the *Hechalutz* representative from Palestine, in Geneva. In his letter, Shushu reported the positive result and asked for an 'exit', i.e. an escape route to Switzerland and money.[1] Unfortunately, Schwalb, who had only a small staff and limited financial means, could not help him with either. The Westerweel group would have to find its own way to safety.

In the autumn of 1942, Joop Westerweel reconnoitered a route over the Belgian border at Budel, a village to the south of Eindhoven. This route was set up by Jasper Daams, who worked at Philips in Eindhoven, where the headquarters of this company were situated. He came from Loosdrecht and knew Mirjam's father, Barend Waterman, who had a partner in Antwerp in connection with his diamond business. At Waterman's request, Daams arranged a contact with the partner via Budel and Hamont, a small place just over the Belgian border. This contact went through a brewer's lad who worked in Maarheeze, near Budel, and lived in Hamont.[2] At the request of the *Hechalutz*, Shushu Simon, who because of his clearly Jewish appearance could not function well in public in the Netherlands, took on the task of reconnoitering the route further. Here, he worked closely with Joop and Wil Westerweel, with whom he was good friends.

A Swiss Illusion

Simon's choice of Switzerland as a safe destination was understandable. From the middle of 1940, Jews had tried to flee to this neutral country from the occu-

---

[1] Haim Avni, "The Zionist underground in Holland and France and the escape to Spain," in *Rescue attempts during the Holocaust*, eds. Yisrael Gutman and Efraim Zuroff, (Jerusalem: Yad Vashem, 1974), 555–590.
[2] GFHA, Cat. no. 463, Jasper Daams, Notitie, De Oirschotse Dijk (ca. 1990), 1–3.

pied areas and Germany. From 1941, a motley collection of individuals and small groups in the Netherlands were busy helping these refugees. To get into Switzerland, quite a lot of barriers had to be taken—starting with the Dutch-Belgian border, which was heavily guarded. The control on the Belgian-French border was less rigorous, because northern France fell under the same military authority as Belgium. However, there were German police patrols who thoroughly checked passengers' papers in a wide stretch of land in the north of France, the so called Somme-line. The demarcation line between Vichy France and the occupied part of the country was guarded by the *SiPo* and Vichy gendarmes, both of which did not think much of Jews. After the Germans occupied Vichy France on 11 November 1942, they patrolled the entire Franco-Swiss border. The Germans also used infiltrators to dismantle smuggling organizations for Jews.

According to J. Presser, about 2,500 Dutchmen fled to Switzerland during the war. Just under half of them were Jews, who mostly arrived before mid-August 1942. After this date, the Swiss government, which was afraid of a too large number of aliens and increasing anti-Semitism, closed the border for what it called 'refugees for reasons of race, such as Jews'. After this, more than 20,000 refugees, mainly Jews, had to return to France. Emergency cases such as sick people, the elderly, children and pregnant women were mostly allowed in. Also, several thousands of other refugees, who, for example, were already across the border, were allowed to stay in Switzerland.[3]

When Shushu Simon started his research into a Swiss route in October 1942, the possibility of reaching that country had therefore already become very slim. Based on several sources, especially an account written in 1946 by his wife Adina van Coevorden (they were married in 1942), it is possible to reconstruct their journey, and so to gain insight into the problems faced in finding an escape route.[4]

A Reconnaissance Mission

Shushu and Adina Simon left to go to Belgium via Eindhoven on 2 October. They had false Dutch identification cards and a pass that allowed them to work as foreigners in France. The next day, they crossed the border at Budel by foot, and

---

[3] Presser, *Ondergang*, dl. 2, 36. See also Jean-François Bergier, et al. (ed.), *Switzerland National Socialism and the Second World War, Final Report* (Zürich: Pendo-Verlag, 2002), Chapter 3, 105–176, mainly 111–115. At the end of 1943, Switzerland eased the admission rules for Jews.
[4] GFHA, Cat. no. 26, Sophie/Adina van Coevorden, Typescript: Die Entwicklung des Holl. Hechalutz, 1–44.

Fig. 16: Shushu Simon

spent the night with a trusted farmer. At his advice, they mixed with the churchgoers to walk to the station of Hamont the next day, which was a Sunday. However, they were noticed. A police agent asked for their papers. The French pass appeared not to be valid in Belgium. The police agent did not believe the story that they worked in Paris and were now going back. But they were lucky: the agent warned them of the Germans and let them go.

In the station and the train, Shushu and Adina immediately recognized other Jewish refugees, just as they too were probably recognized. Joop Westerweel had given them the address of some Dutch students in Leuven (Louvain), who were supposed to know a safe route to France. This route was betrayed, however, and no longer existed. So they went to Brussels, where they got a room at friends of Joop. Finally, for a great deal of money, they were able to find a guide, who took them over the French border.

In the train to Paris, the first check went well. The next day, they went to Tours, close to the demarcation line. However, they did not meet the guide who was supposed to take them to Vichy France. Shushu and Adina had to return to Paris, where they had several Jewish contact addresses. They were able to stay for some time with an older married couple who had already helped other

Dutch Jews. They got better papers and the address of a guide who took them to Lyon, in Vichy France, at the end of October.

The unoccupied city was an enormous relief. There were no Germans—and even the street lighting worked. In Lyon, Shushu had discussions with leaders of Zionist organizations, including Marc Jarblum, a leader of the socialist *Poale Zion*. They referred him to Moissac in the southern department of Tarn et Garonne. Here, the *Eclaireurs Israélites de France* (the Jewish-French scouting organization) had a camp for children who had fled from Germany and Poland.

Shushu had several discussions with leaders of this organization and also with leaders of the *Mouvement de la Jeunesse Sioniste* (Zionist youth movement), which had an office in Moissac. However, they did not want to help the Palestine pioneers from the Netherlands. Adina recalled, 'Most people find the plan absurd and they also tell us that they have enough to do with their own people'. Also, a Dutch-Jewish refugee who was staying in Moissac stated that it was impossible to take pioneers to Switzerland or Spain. Everything seemed to have failed, but one of the people present finally said that he was prepared to help.[5]

At the beginning of November, Shushu returned to the Netherlands and Adina continued the reconnaissance mission towards the Swiss border. From Lyon, she travelled by train and bus to Thonon at the Lake of Geneva, near Switzerland. But there it went wrong, because she missed some necessary document. She had a lot of luck, and escaped arrest by the Vichy police.

Adina went back to Lyon and from there to a small place near Clermont-Ferrand. Here, she waited for Shushu, who returned from the Netherlands at the end of November with new plans. Now, besides Switzerland, the plans were also focused on Spain. Shushu also informed Nathan Schwalb, *Hechalutz's* man in Geneva, of these new plans. The plans concerned a proposal to the Dutch consul in Perpignan to smuggle a whole group of Palestine pioneers who were staying in the Netherlands to Spain. From there, they would have to go to Libya to serve in the British army. Shushu's proposal was controversial and had already led to a sharp disagreement with the pacifist Joop Westerweel. He had threatened to stop cooperating if the plan were to go through.[6] However, the consul (almost certainly Joop Kolkman who had already helped many Dutch refugees) did not accept Shushu's proposal.[7] He was possibly afraid of a German provocation.

After this, Shushu Simon went to Grenoble to speak with Otto Giniewski, a medical researcher at the local university and Jewish resistance leader. It is likely

---

[5] Ibid., 36.
[6] GFHA, Cat. no. 84, Verklaring Arthur Heinrich, 3.
[7] Sierk Plantinga, "Joseph William Kolkman (1896–1944) en de Engelandvaarders," *Negende Jaarboek van het Rijks Instituut voor Oorlogsdocumentatie*, (1998), 10–36.

that Otto's involvement let to much more fruitful ensuing discussions with the *Eclaireurs* in Moissac. It was agreed that Simon would take the Dutch pioneers to the former border between occupied and unoccupied France, where the French would take them over. 'Then, you will be given French maps, food ration cards and also accommodation until you go on to France or Spain. Everything will be accurately worked out and the addresses and roads across the various borders will be written down.'[8]

Shushu Simon left for the Netherlands in mid-December 1942, taking this good news with him. Adina stayed in Lyon, where two other Dutch pioneers, Jopie and Isaac Leuvenberg, joined her. They had bought a flight to Switzerland for 'a large sum of money', but their helpers had abandoned them in Paris. In the Jewish neighborhood of the Marais, they came across Shushu, who referred them to his wife in Lyon. The three of them decided to try to reach Switzerland together.

With the help of the Jewish resistance in Lyon, a guide was found in a border town. He would take them to the Swiss border on Christmas Eve, when little control was expected. After a journey full of problems, whereby they only just escaped (permanent) arrest with a great deal of luck, they reached their destination near the Lake of Geneva. A drunken and loudly-singing guide took them to the border, where the Swiss police were waiting. Because the Leuvenbergs had family in Switzerland who had fled earlier and Jopie was able to show a (false) attestation of pregnancy—and possibly also because it was Christmas— they were not sent back but allowed to stay.[9]

Shushu Simon had less luck. Together with Joop Westerweel, he organized the flights of two groups of pioneers (one group with two people, and the other, with four) from the Netherlands to Brussels, according to the plans that had been made earlier. Shushu was arrested, however, at the end of January 1943 near the Belgian border. The Germans took him to the prison of Breda, where he committed suicide on 27 January 1943, afraid that he would not be able to withstand the heavy interrogations and possibly also exhausted by continually travelling in dangerous circumstances.

According to the Israeli historian Haim Avni, who researched the escape routes, Simon had little success during his three-month-long search.[10] However, this conclusion appears too negative. The experiences of Adina and several other pioneers clearly showed that the closure of the border to Jews and the occu-

---

[8] GFHA, Cat. no. 26, Van Coevorden, Typescript: Die Entwicklung, 38.
[9] Jochebed Leuvenberg-Nathans, *Twee Palestina-pioniers in oorlogstijd* (Winsum: Profiel, 2000), 30–50. The descriptions of Adina Simon and Jochebed Leuvenberg largely correspond with each other.
[10] Avni, "The Zionist underground," 560.

pation of Vichy France all but put an end to Switzerland as a place of refuge. The countless controls in the area made it difficult to reach the border. Also, once they arrived, the refugees were dependent on the Swiss border guards, who checked everybody strictly and turned many of them away. Moreover, the route appeared to be sown with traitors and profiteers, who saw Jews only as easy targets.

Once she was in Switzerland, Adina Simon realized that Schwalb and the *Hechalutz* were not able to do much to help refugees. They were closely watched by the Swiss police and did not dare to take any risks, because there was a large chance that their work would be made completely impossible.[11] Adina did what she could and acted in Switzerland as a sort of communication point between the Dutch *Hechalutz* and the pioneers at various places outside the Netherlands.

Initially, the negligible help offered by the Jewish organizations in Vichy France was disappointing. At the end of 1942, the *Hechalutz* board gave *Werkdorp* pioneer Kurt Benjamin (who, from the beginning of the 'hiding operation', had urged the departure of the pioneers from the Netherlands) permission to research, together with Simon, the possibilities of a flight to Spain. At the beginning of 1943, when he arrived in France, he ascertained that the French Zionist youth movement was not as well organized as the Dutch *Hechalutz*. According to Benjamin, the French Jews in the recently occupied part of the country had no idea what a German occupation meant. In Lyon, the Jews lived 'in an atmosphere of false peace and did not foresee the danger that hung above their heads.' Very soon afterwards, the Germans carried out raids in Lyon, and the peace disappeared.[12]

The poor understanding they had of what a German occupation could mean, was certainly part of the reason the French Zionists resisted the escape plans. In addition, it was much easier to hide in sparsely populated France than in the Netherlands. Moreover, French red tape could be easily circumvented with false papers, which were widely available. Basically, the French Zionists initially had no clear picture of why a dangerous journey to Spain should be undertaken. Political contrasts possibly also played a part in the difficult cooperation. In the *Eclaireurs* (the Jewish scouts), the right-nationalistic youth organization *Beitar* had quite a lot of influence. *Beitar* would have been skeptical about working with the socialist *chalutzim*.[13]

---

[11] NIOD, Doc. II, 614 (A+B), Adina Kochba and Rina Klimov (eds.). Translation Channa Hoffman, Manuscript: Het verzet van de Nederlandse Chaloetsbeweging en de Westerweelgroep, bijdrage Adina Simon, hfdst. 7, 9, 10. (From now on: Kochba and Klimov, Manuscript).
[12] Kochba and Klimov, Manuscript, Bijdrage Kurt Benjamin, hfdst. 8, 9 – 11.
[13] GFHA, Cat. no. 48, Hans Ehrlich, Brief aan Kurt Benjamin, 23 maart 1990.

In carrying out all the activities of the Dutch pioneers in Belgium and France, it was obvious how important money was. With money, it was possible to operate more safely and pay reliable guides, bribe civil servants and purchase extra documents. Mr. and Mrs. Leuvenberg-Nathans could go to Paris without any problems. Later, they could buy a false pregnancy attestation, which helped with their entry into Switzerland. Adina and Shushu Simon and other pioneers often ran into problems at the Belgian border. Money, which the pioneers lacked, was no guarantee for a safe journey, but it could significantly reduce the risks.

A Route to Spain?

At the beginning of 1943, Shushu Simon and the *Hechalutz* leaders in the Netherlands were convinced that Spain, instead of Switzerland, was the preferable destination. Putting the plans together, however, was easier said than done. Spain posed the same problem as Switzerland: namely, how could the pioneers get there safely?

During the Spanish civil war (1936–1939), Spain's leader, the right-wing nationalist General Franco, had received a great deal of help from Germany and Italy. After the capitulation of the Republican troops in the spring of 1939, Franco decided to first consolidate his regime. His sympathy, however, lay clearly with the Axis powers. This was also expressed in his regime's attitude to the approximately 6,000 Jews who lived in Spain. Their addresses were collected, and through the officials of the fascist Falange there was official contact with the SS in Berlin about their persecution. In 1941, Franco allowed a Spanish division of volunteers to go and fight with the Germans on the Eastern Front.

In the chaotic period after the French capitulation in July 1940, Spain admitted thousands of refugees, including many Jews. Some of them entered the country with transit visas that were issued by the consulate in Bordeaux and others without prior knowledge of the Spanish government. Jewish refugees were repeatedly sent back in both 1940 and afterwards. One known example is the writer Walter Benjamin, who committed suicide in September 1940, as a consequence of fearing deportation. There was no clear policy in sending people back; it depended mostly on arbitrary decisions taken by local commanding officers.

After the Allied landings in North Africa in November 1942, which signaled the defeat of the Germans and Italians there, a conservative nationalistic movement in General Franco's government gained more influence at the cost of the Falange. From the Spring of 1943, refugees were offered sanctuary in Spain and treated reasonably well.[14] However, the Pyrenees remained a formidable barrier, which was almost impassable from the early autumn to late spring because of snow and cold. Moreover, a 40-kilometer-wide stroke of land (later in the war, 80 kilometers wide) on the French side of the Pyrenees was forbidden for non-inhabitants. The small flat parts along the coasts of the Mediterranean and the Bay of Biscay were heavily guarded.

In mid-January 1943, four pioneers from the *Werkdorp Wieringermeer* and two from Loosdrecht left in pairs, shortly after each other, to go to Spain. The first pair consisted of Werkdorpers Kurt Benjamin and Kurt Ehrlich, with Ehrlich having escaped from Westerbork shortly before. Joop Westerweel took them across the border at Budel and accompanied the two to Brussels. Here Shushu Simon took charge and travelled with Benjamin and Ehrlich to Paris. A 'guide' then took them over the former demarcation line, which was still guarded.

Benjamin and Ehrlich, who looked non-Jewish, had false identification cards. They said they were Dutch workers, who were returning to France from a holiday. The two were arrested in both Belgium and France but talked their way out of it. In Lyon, they reported to the *Hechalutz*, according to Shushu's instructions. However, after a week, they were told that there was no possibility of going to Spain. They were given (badly) forged French papers, after which a French pioneer took them to Nice, which had been annexed by Italy. They were relatively safe there because the Italians did not actively persecute Jews.

Benjamin and Ehrlich stayed in Nice for about two months. During this time, they concluded 'that the movement in France was not capable of helping us, except with material support'. They then looked for work on a farm in the surroundings and were given good papers by Dutchmen in Nice. When the Badoglio government deposed Mussolini in August 1943, and signed a peace treaty with the Allies, German troops occupied northern Italy, including Nice and its environs. The two then went to Toulouse, and from there partly by foot to the small village of Saint-Girons, at the foot of the Pyrenees. A first attempt to go to Spain failed, but a second attempt at another place was successful. Accompanied by a guide, Benjamin and Ehrlich (who had joined a group of eight Jews

---

14 Stanley Payne, *Franco and Hitler* (New Haven/London: Yale University Press, 2008) mainly 210–220. See also Michael Alpert, "Spain and the Jews in the Second World War," *Jewish Historical Review*, vol. 42, (2009): 201–210. It is assumed that between 25,000 and 30,000 Jews entered Spain during the Second World War. However, Spain did virtually nothing for Sephardic Jews in Greece and elsewhere, who had Spanish passports. It was not until 1944 that 1600 Sephardic Jews were allowed into Spanish Morocco.

and non-Jews) reached Catalonia at the beginning of September 1943—after a 'very heavy journey' of four days.[15]

The second pair consisted of *Werkdorp* pioneers Mosche Osterer and Fritz Bachrach. Osterer came from Austria, was 22 years old and arrived in the Netherlands at the beginning 1939. After the *Werkdorp* was closed, he worked for some time in a vegetable garden project in Amsterdam. He then went to work as a gardener in *Het Apeldoornse Bos*. No more is known about Bachrach, other than that he participated in the individual *hachshara* for some time from Deventer. They both had false papers and were taken to the border near Budel by Norbert Klein on 21 January 1943. From there, they travelled on to Brussels, where they met Shushu the next day. He gave them Belgian ration cards and money and told them to wait in Brussels for a few days until he was back from the Netherlands.

Osterer and Bachrach waited nine days for Shushu—but unknown to them, he had been arrested in the meantime at the Belgian border. Because their money ran out, they travelled back north to a previously-agreed emergency address in Antwerp. Here, they received a telephone call from Joop Westerweel, who told them to return to Brussels and wait for him there. They then travelled together to Antwerp again, where they met the third pair of people trying to reach Spain. These were the 'Loosdrechters' Willy Gerler and Heinz Sechestower, just 19 and 18 years old, respectively. They were accompanied by a Belgian guide 'Rik', about whom there is more, later.

The four of them travelled with Rik to Paris by train—and from there, immediately on to Bordeaux, where 'Rik' handed them over after a few days to a French guide. They again went by train to a small place near the Spanish border, via Bayonne. Here, they dressed up as tourists, after which the guide, who did not dare to go with them, outlined the rest of the route. 'My mood was fatalistic, Fritz B. was quite excited and the two young men were in high spirits', Osterer recalled. The four of them walked through a hilly landscape for three hours and arrived at a Spanish café. The owner was astonished to see them. The place where they were crossed the border was always guarded by two German military posts.[16]

These, then, were two successful escapes to Spain. They were so unusual, however, that they had very little significance in terms of providing information

---

15 NIOD Doc. II, 614 (A+B), Kochba en Klimov, Manuscript, Bijdrage Kurt Benjamin, hfdst. 8, 9, 10.
16 GFHA, Cat. no. 166, Verklaring Moshe Osterer, december 1955, 1–4. See also GFHA, Cat. no. 22, Interview Mirjam Pinkhof-Willie Gerler and Peter Tower (Heinz Sechestower), 27 april 1994. Gerler and Tower both lived in the United States.

Fig. 17: Willy Gerler shown here during his hachshara training in Loosdrecht. He was one of the first pioneers who reached Spain at the beginning of 1943.

for taking others across the border. Benjamin and Ehrlich had taken eight months to arrive in Spain and were disappointed that the help promised by the Jewish resistance in France had at first been so limited. It is not clear whether they had much contact with the Dutch *Hechalutz* while they stayed in southern France. The Jewish resistance, which became somewhat better organized, did help them later with their journey over the Pyrenees.

Partly considering the speed at which the plan was set in motion, the escape of the four pioneers to the Atlantic coast was a success. However, it was a success that was based more on luck than on knowledge of the situation. For this reason, it was of doubtful importance as an escape route for other pioneers. Mosche Osterer and the three others were arrested almost immediately after crossing the Spanish border—but not sent back, as they had feared. They were kept prisoners for a short time, but released as a result of the efforts of the British consul. The four were then subjected to several weeks of house arrest and three months imprisonment in an internment camp. They were, however, able to send a coded message to the Netherlands about their safe arrival.[17]

Kurt Reilinger and the New Escape Route

After Shushu Simon's death, the *Hechalutz* board gave 25-year-old Kurt Reilinger the task of working out the escape route to France and Spain. Reilinger was born into a middle-class family in Stuttgart at the end of 1917. His father, a salesman, was Jewish; his mother was Catholic. Kurt and a younger brother were active from an early age in Jewish youth organizations—the last of which was a chapter of *Werkleute* (working people), a group that was allied to the left-wing socialist *Hashomer Hatzair*. Despite poor health, Kurt had various management positions in the organization.

After secondary school, Reilinger, a serious young man with a great sense of responsibility, went to Mannheim to work for *Werkleute* as a youth leader. In Mannheim he also followed a course to become a metal worker. After a year, he was also made youth leader in Frankfurt. In 1939, he went to Palestine with his group, but he returned to Germany after a month and a half to lead a *hachshara* course near Berlin, as he had promised. Shortly before September 1939, he went to the Netherlands, where he went to work in the *Wieringermeer* and was also a board member of the *Hechalutz*. After the *Werkdorp* was closed, he went on individual *hachshara* with a farmer near Deventer.[18]

When Reilinger was asked in February or March of 1943 to succeed Shushu Simon, he was hesitant. He had a more introverted personality than Simon and had spent the past year and a half or more on a farm—far from the pressure of the persecution. He was of course aware of this pressure, through his management position. An important advantage was that, as a half-Jew, he could move about more easily without a star on his sleeve and without a J-stamp in his papers.

Reilinger's hesitation would possibly partly have been caused by the question of whether he thought he was the right man to solve the great problems that existed at the time.

In the first place, the contacts with the Zionist movement in France, which had not been very good, and had been broken by Simon's death.

More seriously, there was pressure to find hiding places, especially for the Deventer group pioneers, which increased strongly in the beginning of 1943.

---

17 GFHA, Cat. no. 22, Interview Mirjam Waterman-Gerler and Tower.
18 Kochba and Klimov, Manuscript, Bijdrage Roda Reilinger, hfdst. 7, 45, 46.

This was because many temporary exemptions for reporting to Westerbork were withdrawn at this time, while there were insufficient hiding places in the Netherlands.

Kurt Reilinger probably decided in consultation with Joop Westerweel to further investigate the Bordeaux route, which had proven to be successful. It is not clear if the rather fortunate manner in which Osterer and his three comrades had crossed the border was known in the Netherlands. The size of the groups that went to France increased considerably at this time, against the wishes of Westerweel. Reilinger also sought contact with both (non-pioneer) Jews and non-Jews who wanted to go to France.

Fig. 18: Kurt Reilinger took over the leadership of the escape network in France from Shushu Simon in August 1943

The Antwerp Connection

In the meantime, Joop Westerweel continued his search for a good escape route. He had found the route near Budel through Jasper Daams. Daams and his Philips colleague Willem van Heeckeren helped Westerweel create a support point in Eindhoven. Pioneers who went across the border could now stay at this support point for some time.[19] Westerweel had, however, also travelled by train to Antwerp several times to see how professional smugglers worked.

Smuggling had always been a popular activity in the border region between the Netherlands and Belgium. All products that were treated differently in customs and excises were potential contraband. In the 1930s, for example, this was Belgian margarine. After 1940, smugglers took large quantities of food such as wheat, rice and oatmeal to Belgium. Smuggling in the other direction was dominated by taking French tobacco to the Netherlands. It could sometimes be very rough at the border. German soldiers shot at smugglers, who were sometimes also armed and shot back, to keep them at a distance. The smugglers certainly had a fairly positive image during the war. The population regarded them as clever people— in the style of the legendary literary hero Reynard the Fox— who dared to challenge the Germans. It was found acceptable that the smugglers put their own interests first.

At the end of 1942, Joop Westerweel met Henri 'Rik' Lelièvre, a smuggler who lived in Antwerp. Westerweel probably saw Lelièvre as an intrepid spirit who could be of service in taking Palestine pioneers to safety for payment. Lelièvre was born in December 1900, in Overpelt, a Belgian town just across the Dutch border from Brabant. From 22 to 35 years of age, he worked as a ship's steward. He then for several years had a job as a waiter, until he was called up for enlistment in 1939 because of the threat of war. After the capitulation of the Belgian army he was made a prisoner of war, but together with other Flemish soldiers he was released in June 1940 by the Germans.

Lelièvre went to Germany voluntarily in September 1940. When he returned at the beginning of 1941, he worked in northern France from the end of June. From January to June 1942, he was employed as an interpreter with a German

---

[19] Willem van Heeckeren lived with his family of five children in two luxurious caravans, designed by his wife, on the boundary between Oirschot and Eindhoven. Besides the pioneers in transit, the family also had another two Jews in hiding. These were the baby Susie Weinberg and Thea Perlmutter (born 1924). Lotti Wahrhaftig (born 1926), who came from Loosdrecht and escaped from Westerbork, also lived with the Van Heeckerens for some time, 'Heeckeren Van Family'. https//www.yadvashem., accessed 9 augustus 2014.

company in Bordeaux. Here, he wore a uniform, possibly that of the *Organisation Todt*, and would have had some sort of official German position.

Before he left for Bordeaux, Henri Lelièvre reported to a right-wing Belgian resistance organization, the *Nationale Koningsgezinde Beweging* (NKB, National Royalist Movement). After his return, he took 'people in hiding and resistance people' to France several times in September and October 1942. At his commander's order, he could also have accompanied several English pilots, who had been shot down and had arrived from the Netherlands, to Compiègne in northern France.

Lelièvre's personal life was just as colorful as his professional life. He was married and had two children. However, from the beginning of the 1930s, he lived with Julia Cole. She managed 'the inn' *Chez Julia*, which was in fact a brothel in the Antwerp harbor area.[20]

Rik Lelièvre was a daredevil and an opportunist, who would never let a chance pass by to earn some money, no matter the risk. It is obvious that he would have told Joop Westerweel about his journeys to France for the Belgian resistance. It is likely that he was less forthcoming about his other activities. The two then came to an agreement. According to a post-war statement by Maria Cole, the sister of Lelièvre's girlfriend, 'one and the same man, a Dutchman, took two to three Jews each time'. The man, almost certainly Westerweel, 'paid him two thousand francs for each man that he took across the border each time.' The Jews stayed in the inn and Julia Cole also received a payment for this.

Maria Cole travelled to Bordeaux twice with Lelièvre. 'We took three Jews each time and when we arrived there, I had to wait in a café while Rik took the Jews away. I never knew or saw who he took them to. Besides these two times that I accompanied Rik, he took Jews to France on his own.'[21] It is not clear how many Jews Lelièvre accompanied. Maria said she thought there were nine, which definitely included Mosche Osterer and his three comrades. Lelièvre possibly also transferred Jews other than Palestine pioneers. It is not known whether they all arrived in Spain.

According to Maria, three other Jews followed later. Two of them were pioneers, but they did not get far. On the eighteenth of March 1943, Lelièvre and his fellow travelers were arrested in Compiègne, in northern French. His name appeared to be in a search register as a currency smuggler. The Germans took

---

20 Soma-Ceges, Archief Dienst voor Oorlogsslachtoffers, Brussels Belgium. Diverse documenten betreffende H.A.M.G. Lelièvre, vooral: Verslag 7 oktober 1951 met daarin Verklaring van J-E. de Ridder, provinciaal bevelhebber weerstandsgroepering Nationale Koningsgezinde Beweging (NKB) en rapport 21 februari 1953.
21 Ibid., Verklaring Maria Cole, in Verslag.

him to the Sachsenhausen concentration camp near Berlin. In the autumn of 1944, he was transported to Neuengamme, where he died at the end of March 1945.[22] The two pioneers, of whom the identity is not known, were probably sent to the east via Drancy. For some time after Rik Lelièvre's arrest, another 13 pioneers made their way to Julia Cole's inn. There is more about this later, under the heading Adventures in Antwerp.

The curious association between Westerweel and Lelièvre fitted into the former's romantic-anarchistic worldview, which recognized a sort of kindred spirit in Lelièvre. By the way, there is no evidence that the Antwerp smuggler was involved in any treason. It is unclear why the German made the accusation of currency smuggling; Rik Lelièvre might have taken money to Bordeaux for private individuals or the resistance, and the Germans discovered this somehow.

Finding a safe escape route to Switzerland and later Spain was a difficult matter that the Westerweel group—mainly Shushu Simon—had to figure out on their own. The betrayal of the eight pioneers in October 1942 was an important factor here. After this betrayal, the group's leaders decided to cease working with others. But 'doing it on their own' was also important in the socially marginalized position of the group's Jewish and non-Jewish members alike. Joop Westerweel's old contacts in Belgium no longer functioned and the Palestine pioneers found only a limited degree of connection with the Zionist movement in France, where the *Hechalutz* was not strongly developed. Shushu Simon was therefore left to depend on contacts with official Zionist circles and the Jewish scouts, who probably regarded his enthusiastic left-wing idealism as somewhat strange.

What could have come from cooperation with the Jewish resistance in France, which was promised at the end of 1942, is very much the question. Considering Kurt Benjamin's experiences, this was probably not much. The Jewish resistance in France was not organized in such a way that it was capable of undertaking crossings to Spain until mid-1943.

The choice to work with the Belgian Rik Lelièvre and his girlfriend Julia Cole was characteristic for the isolation in which the Westerweel group operated. Joop Westerweel thought that he had met a kindred spirit, a working-class anarchist, in Lelièvre. However, money appears to have been the main motive for the Antwerp smuggler. Here, it must be noted that Lelièvre put himself in great danger for relatively small amounts of money, and this finally proved fatal for him.

---

22 Ibid., Verklaring P.M.H. Stevens, 17 september 1949.

8 Hiding in Limburg, Germany and France

The first months of 1943 formed a difficult period for the Westerweel group. Shushu Simon and Rik Lelièvre, the guide to Spain, had been arrested, so that the possibility of an escape route to Spain had to be abandoned for the time being. It took Kurt Reilinger, Simon's successor, several months to restore the connection with the Jewish resistance in France. The pressure on the group increased, because many of the stamps offering exemption from reporting for transport to Westerbork of Palestine pioneers were withdrawn. These pioneers often also asked the group for help to find hiding places.

Chiel Salomé and Sevenum

However, there was also a positive development in this period. In the spring of 1943, Joop Westerweel came into contact with Chiel Salomé, a civil servant of *Rijkswaterstaat* (Department of Waterways and Public Works) who was born in Kortgene in the province of Zeeland in 1912. For his surname he thanked his Huguenot family, which left France at the beginning of the eighteenth century and immigrated to Zeeland to escape the marginalization of the Protestants. The Salomé family had eleven children. His father, a flour trader, came from a miller's family and had also worked as a miller's lad in Zutphen for some years.[1] Chiel Salomé was an individualist spurred on by the realization that the anti-Semitic measures of the occupier were unjust and that they had to be resisted.

At the end of the 1930s and the beginning of the 1940s, Salomé worked as a supervisor second class at the *Rijkswaterstaat* in Hardinxveld. He shared a room there with a Jewish colleague, Bendiks, with whom he often debated political matters and the situation in Germany. Through Bendiks, he also met other Jews from Gorinchem and Hardinxveld. In the summer of 1942, when the first calls came to report to Westerbork, Chiel found a hiding place for Bendiks and eleven Jews from Gorcum.

He then sought contact with what he called a 'communist resistance group' in Amsterdam, for whom he took Jews to addresses to hide in the weekends. During this period, Chiel sometimes worked together with his brother Henk to bring Jews to villages in the northern part of the southern province of Limburg. In mid-December 1942, he was arrested at the Amsterdam Central Station, together with

---

[1] E-mails van J. van Haver and G. de Fouw, gemeente Noord-Beveland, aan auteur, 4 en 9 april 2013.

a Jewish woman who he was accompanying to a hiding address. The *SiPo* interrogated him several times and then held him prisoner for six weeks in the Amsterdam *Huis van Bewaring* (Detention Center).[2]

Chiel Salomé resumed his resistance work almost immediately after his release in February 1943. After the arrest of two Jews whom he had helped, however, he was warned that the Germans were looking for him again. His landlady gave him an address in the province of North Brabant, where he hid for three months.

It was during this period that Chiel connected with the Westerweel group through a Rotterdam acquaintance of his mother who worked at the Montessori school on the Beukelsdijk. At Joop's request, he renewed his relation with Limburg resistance organizations, where he had earlier taken Jews to go into hiding. Two resistance people who he knew well from Schaesberg (in the southern part of Limburg) mentioned the name of 'Miss Eugénie' Boutet. She was the principal of the Sevenum primary school for girls. Salomé then travelled to Sevenum, in the northern part of Limburg, with a Jewish married couple. He waited until lunchtime to speak with the school principal, to discuss the possibility of hiding Jews with her. Boutet told Salomé to leave the couple behind with her in Sevenum. She referred him also to Pastor Vullinghs of Grubbenvorst, with whom he made an appointment. Fourteen days later, 'Miss Eugénie' sent him a message: 'You can bring anybody you want'.[3]

Several times in the summer and autumn of 1943, Chiel Salomé accompanied Jews by train to Venray, after which they walked six kilometers to Sevenum. Eugénie Boutet would take them in charge at that point, providing further transport to the definitive hiding address. This was often in Sevenum, but sometimes also in other places in northern and central Limburg. The 'Loosdrecht' sisters Gustel and Sophie Nussbaum, who were mentioned earlier in their failed attempt to hide in Amsterdam (see the chapter, The dangers of being in hiding), also belonged to this group.

In taking the people in hiding away, Salomé followed the orders of Joop Westerweel. For safety reasons, Salomé usually did not know who these people were or what their background was. He took Jews not only to Sevenum, but also to addresses in south Limburg and Friesland (in the north of the Netherlands),

---

[2] GFHA, Cat. no. 201, Verklaring Machiel (Chiel) Jan Salomé, 1955 en november 1989. In 1955, Chiel Salomé made a fairly superficial statement about his resistance activities, in which the Westerweel group is not mentioned. His statement of 1989 is much more detailed, but the report, which was probably written by Mirjam Pinkhof in a sort of telegram style, contains several (chronological) inaccuracies.

[3] Ibid.

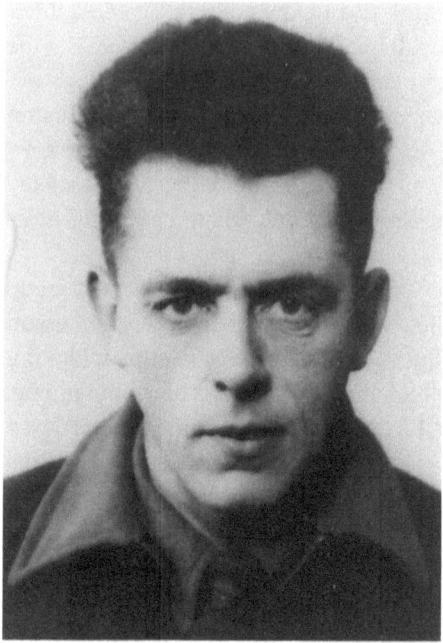

Fig. 19: Chiel Salomé took pioneers and other Jewish people who wanted to hide to Limburg and Friesland

with Sneek as the distribution point; once he took them to a farmer in the surroundings of Breda. Sevenum, however, played an unmistakably important part in the hiding activities. Approximately twenty pioneers and an unknown number of other Jews were taken to Sevenum—by Chiel Salomé or by others—and stayed in hiding there for various lengths of time.

Sevenum was a thinly-populated rural community, where only just a few members of the NSB lived. Around 1900, this agrarian region had developed further through the excavation of bogs and the introduction of horticulture, with a large export to Germany.[4] The Netherlands has many such villages as Sevenum. The inhabitants of most of these small villages ignored the war as best they could. However, the largely Catholic Sevenum had a priest with a keen interest in the political situation, who was active right from the very beginning of the war in the resistance against the Germans.

---

4 André de Bruin, *Noord-Limburg integraal bekeken, 1850–1950* (Sittard: Mooi Limburgs Boekenfonds, 2010), 147–154.

A Catholic Non-Conformist

Father Henri Vullinghs was the son of a local brewer who was born in Sevenum in 1883. He studied musicology in Rome and New York in the 1920s. In New York, he followed a course from the musicologist Justine Ward, who had developed a method to bring new life to Gregorian choral singing. In 1928, he was given the position of director of the Dutch Ward Institute in Roermond. Here, the tutors taught the method to Catholic teachers. The year after this institute was made independent in 1938, Vullinghs was appointed as priest to Grubbenvorst, a village close to Sevenum.[5]

The new pastor was a conservative Catholic, but also an erudite and tolerant man, who had developed an aversion for fascism during his study in Rome. This applied equally to German National Socialism, against which he regularly warned in the 1930s. Through his position in the Ward Institute, he had many contacts in the music world. Because there was also much choral singing in SDAP circles, his circle of acquaintances also included social democrats, some of whom were of Jewish origin. In doing so, Father Vullinghs explicitly ignored a 1924 decree issued by the bishop of Limburg to avoid contacts with Jews. This was an unusually open attitude for a Catholic pastor in the Limburg of the 1930s.

Immediately after the capitulation in May 1940, Vullinghs gave a sermon in which he said that the Dutch army had lost the struggle, but that the war was continuing. From the summer of 1940, he kept himself busy giving refuge and helping escaped French prisoners of war. In this, he was assisted by his two chaplains. Also Eugénie Boutet, the principal of the girls' school, who was 45 years old in 1942, was of great importance in efforts to provide refuge to the French. She had family in Limburg, but originally came from Luik (Liège) in Belgium and spoke fluent French.

Later in the war, Vullinghs and his helpers also helped crews of Allied airplanes who had been shot down and wanted to return to England via France and Spain. When the transports to Westerbork started in 1942, Vullinghs also helped Jews. In this endeavor Vullinghs worked together with Mathieu Smedts, a socialist journalist friend of his, who came from Sevenum and lived in Amsterdam.[6]

---

5 Wilko Brouwers, "Ward in Nederland, toen and nu," https//www.wardcentrumnederland.eu., accessed 3 juni 2013.
6 Cammaert, *Het verborgen front*, 136–144. Herman van Rens, *Vervolgd in Limburg* (Hilversum: Verloren, 2013), 222–226. See also De Bruin, *Noord-Limburg*, 218–224.

Only a few of the Loosdrecht group, such as Sophie Nussbaum and later Ruth Tischler, went into hiding in the Sevenum area. Another 15 pioneers from the *Werkdorp* and the Deventer group found hiding places there. The number of 'about 30' mentioned by Cammaert appears too high, as certainly does the estimate of 'between 75 and 100' mentioned by Van Rens.[7] Lily Kettner, who was 20 years old and born in Vienna, had escaped shortly before from Westerbork. She arrived in Sevenum in September or October of 1943. Chiel Salomé took her there, partly because of her dark hair and German accent, which was less conspicuous in Limburg than elsewhere. Lily, who made fairly extensive reports (in 1957 and 1987) of her stay, hid for about a year and a half in Jan and Nelly van Enckevort's farm, where she was comfortable: 'I joined in all sorts of work such as domestic work and work on the field and earned the wage of a maid. The money went to our organization (the *Hechalutz*) ...'

Sophie Nussbaum who, as described, had arrived in December 1943, also spoke kindly about the people who hid them and about Sevenum. They lived on a small farm. However, she did have to go to church every Sunday, while her host continually tried to convert her. As a sort of defence, Sophie described in detail 'the Land of Israel, and found a great deal of understanding'. However, he could scarcely believe that Jews wanted to work there as farmers.[8]

Pioneers in Northern Limburg

The true extent of the efforts in Sevenum was clearly demonstrated on one occasion to Harry Asscher, religious helper of the *Hechalutz*. After he arrived in Sevenum with pioneer Klara Rothenberg, Asscher went to borrow a bicycle from an acquaintance because of the great distance to her hiding place. When he returned, Klara had disappeared. He became very anxious, but was told by an acquaintance: 'Everything is fine. The police have brought her to the police station, for her own safety'.[9]

This positive attitude was without doubt fostered by the influence of Vullinghs and his assistants. The priest of Grubbenvorst totally ignored a prohibition from his bishop to be involved with the resistance. Kettner estimated that Sevenum, a place with some 3300 inhabitants, must have had some 300 people of all

---

7 Cammaert, *Het verborgen front*. See also Van Rens, *Vervolgd in Limburg*, 224.
8 GFHA, Cat. no. 161, Verklaring Sophie Nussbaum, 2.
9 GFHA, Cat. no. 9, Verklaring Harry Asscher, waarschijnlijk augustus 1955. Rothenberg could only stay in Sevenum for a short time.

sorts in hiding. According to Eugénie Boutet, of these people, there were initially around 80 and later about 100 Jews.[10]

At the beginning of 1944, about ten of the Jews in hiding were from the *chalutzim*. Among them was Maurits Waterman, the younger brother of Miriam. The pioneers met on dark nights in a chicken coop near the house where one of them, Walter Posnanski, was in hiding. Joop Westerweel was present at one of these meetings. His wife Wil sometimes also took money, ration cards and other items to northern Limburg. She reported there to a milliner in Velden near Venlo. 'There, someone came to pick me up and took me over the river Maas (Meuse) to Grubbenvorst in a rowboat ....'[11] After the arrest of Chiel and Wil in December 1943, Elly Waterman and Harry Asscher assumed the care for the pioneers.

Northern Limburg was certainly no paradise for Jews in hiding. Rumors about their stay, the careless behavior of several of them and acts of betrayal led, in the spring of 1944, to raids by the *SiPo/SD* and a special unit of the Dutch police from Eindhoven. Father Vullinghs was arrested at the beginning of May 1944 and taken to Camp Vught. Later, he was sent to the concentration camps of Sachsenhausen and Bergen-Belsen, where he died in April 1945. Eugénie Boutet was also forced to flee at the beginning of May 1944. She finally found accommodation in a nearby village. Other resistance people from northern Limburg had to go into hiding as well. Ten Jews, none of them pioneers, were arrested in the raids.

In the summer of 1944, Sophie Nussbaum received an anonymous letter, 'which accused me of being a Jewess'. Shortly before, a local government civil servant had given Sophie a properly forged identification card, which stated that she was an inhabitant of Sevenum. 'However, when a second anonymous letter came, in which I was threatened with police action, I left the place.'[12]

Sophie was housed in Voorst, Gelderland. She did not feel safe there and decided to return to Sevenum. Because of the railway strike that had started on 17 September, she walked to Nijmegen. The next day she went to Roermond, where her sister was in hiding. However, she could not stay there and went on to Sevenum over the Maas bridges. Here, she again found a hiding place and was liberated by the English army at the end of November 1944.

For Lily Kettner as well, the final months of the occupation were accompanied by great tensions. Because of the danger of betrayal, she and Ruth Tischler,

---

10 GFHA, Cat. no. 400, Interview Mirjam Pinkhof-Eugénie Boutet, 1 juni 1991, 4, 5. Van Rens mentions comparable numbers, *Vervolgd in Limburg*, 225, 226.
11 Westerweel, *Lijn of cirkel*, 70.
12 GFHA, Cat. no. 161, Verklaring Sophie Nussbaum, 3.

an 18 year-old fellow pioneer from Loosdrecht, were to go to an address to hide that was outside the village of Sevenum. However, because northern Limburg had become a war zone through the Allied advance, this was not possible.

Shortly afterwards, a German officer came to the farm they were staying at, to say that 30 soldiers would be quartered there to recover from the fighting at the front. During this time, a new group of soldiers arrived every few days. The farmers, the Van Enckevort family, kept the young women with them as helpers with the chores—Lily, who knew farm work, was the 'hired girl' and Ruth was 'her cousin from the city'. The two slept in the stable above the cows and agreed 'to always be together, so that the soldiers would never find us alone.'[13] This went well until the liberation.

Hiding in the *Arbeidseinsatz*

In the spring of 1943, an escape route cropped up that at first sight seemed less viable, but had certain advantages just for this reason. A civil servant of the regional employment office in Enschede told the leaders of the *Hechalutz* about the possibility of letting pioneers hide in Germany as Dutch workers who were sent there as part of the *Arbeidseinsatz* (forced labor internment). Through the *Hechalutz*, two groups of five pioneers each, probably all from the Deventer group, were sent from the employment office to Bielefeld and Dortmund with false identification cards and papers. All pioneers had followed professional training or had worked in agriculture.

The group to Bielefeld left first. When everything appeared to go well, the pioneers for Dortmund followed a month and a half later. Contact between Amsterdam and both groups was quickly broken, however. By coincidence, several other pioneers went to work in Germany several months later when, as described before, they were arrested at the Dutch-Belgian border. Under interrogation they were able to keep to their story (which was based on false papers) that they were Dutch workers and illegally on their way to France. They were then sent to the *Arbeitseinsatz* in Germany.

One of them, Hans Mogendorff, arrived in Essen at the end of July 1943, but got caught in a heavy bombardment the first night, whereby he lost his possessions. As compensation, he received 100 marks and a week's leave in the Netherlands. Mogendorff had been a *Hechalutz* board member for some time before the war. He had good contacts with Hannemann and Reilinger, whom he immediate-

---

13 GFHA, Cat. no. 111, Interview Mirjam Pinkhof-Lily Kettner, 1987.

ly visited in Amsterdam. They asked him to return to Germany to try to restore contact with the pioneers in Bielefeld and Dortmund.[14]

Mogendorff did this, but the pioneers in Bielefeld seemed to have disappeared, and were probably arrested. The group in Dortmund was still there. On the basis of his own experiences, Mogendorff had in the meantime discovered how the arrests had gone. The German police required that the Dutch workers in Germany had certain passes; to issue these, they asked for information from their identification cards. In the case of the pioneers, these documents were often stolen or 'lost'. The numbers on the cards were on a 'black list' in the Central Population Register in The Hague. The excellently organized register, manned by Dutch government officials, passed this information along to the German police, after which the people involved were arrested. These procedures took about three months, which is why the Hechalutz board initially thought that going into hiding in the *Arbeitseinsatz* worked well.

When Gideon Drach and Norbert Klein were arrested in the Netherlands, a list with names of the people in hiding in the *Arbeitseinsatz* fell into German hands. Consequently, the group in Dortmund had to leave immediately. They were helped by several Dutch border workers, who lent them a special pass to cross the border safely. Hans Mogendorff was able to warn two other pioneers, Heinz Moses and Heinz Levin, in time. These pioneers had also been sent to Germany at the same time. Sometime later, when he wanted to return to the Netherlands, Mogendorff was arrested at the border, heavily interrogated and mistreated. He was then given a six-month prison sentence because of illegally crossing the border. He did not return to the Netherlands until April 1944, after which he was able to hide near Deventer.[15]

Hiding in the *Arbeitseinsatz* probably cost five of the ten *chalutzim* their lives. Greet Stork, the girlfriend of Heinz Herz, one of the pioneers who had not returned, wondered after the war if it had not been irresponsible to send them to Germany without good papers. It was indeed a risky solution, which can be explained by the great pressure under which the *Hechalutz* leaders had to operate in the spring of 1943. There were insufficient hiding places in the Netherlands and no safe route to France. The Dutch civil servant who thought up the idea would not have realized that the identification card information would be checked again in Germany. It is likely that more Palestine pioneers went into hid-

---

14 Mageen (Mogendorff), *Van zonsondergang*, 31–40.
15 Mageen, *ibid.*, 41–48. There are several versions of the German adventure, including a romanticised historiography of Heinz Moses. Also, the number of arrestees varies between four to seven or eight. See also GFHA, Cat. no. 153, Manuscript by Heinz Moses, written in 1944 before he went to Palestine. However, Mogendorff's version appears to be the most reliable.

ing in the *Arbeitseinsatz*, but then at their own initiative. For example, the Gouda pioneer/writer Jakov Lind worked as a sailor on German inland vessels from November 1943 and later in a laboratory in Berlin. He survived the war.[16]

Adventures in Antwerp

It took some time before the news reached the Netherlands about the arrest in mid-March 1943 of the Antwerp people smuggler Rik Lelièvre. In April, therefore, another two pairs of pioneers left for Spain. The first two were the Austrian brothers Josef and Arthur Heinrich from Loosdrecht. They had been in hiding in IJmuiden, where they were cared for by Jan Smit. He also went with them to Eindhoven, after which Joop Westerweel took them across the border near Budel. There, Lelièvre was supposed to take charge of them and guide them to the Spanish border, just like Osterer's group.

Rik Lelièvre was of course not there; Joop therefore gave the Heinrich brothers some money as well as instructions to wait for him in the Antwerp café *Chez Julia*, of the smuggler's girlfriend. It was not until several weeks later that a card from the Sachsenhausen camp made it clear that Lelièvre had been arrested. In the meantime, the second pair, Werner Kahn and Ernst 'Willy' Hirsch, had arrived in Antwerp. Because their money was almost gone and it was not clear what they had to do when they heard that Lelièvre was arrested, Willy Hirsch (who was 26 years old and somewhat older than most pioneers) travelled back to Amsterdam to ask for both money and advice. He was told that they had to wait for a new group that would also be heading to Spain.

This group of nine men arrived sometime in May. They were taken to the border by 'Theo', Lelièvre's successor. No more is known about this man's identity than that he also came from the raw Antwerp smugglers' milieu. Dressed in a uniform of the *Organisation Todt*, 'Theo' had been the one who had been paid to take Emil Glücker and three Gouda pioneers across the border—which he did, in mid-May 1943, in a German army train. 'Theo' thus made a spectacular entrance in the Westerweel group. According to a post-war statement by Arthur Heinrich, who was involved with him in Antwerp, 'Theo' was an escaped prisoner.[17] With the transfer of the second group of pioneers, there were again some rather strange occurances, as appears from a report by one of the pioneers from Gouda, Berrie Asscher.

---

16 GFHA, Cat. no. 272, Verklaring Greet Stork, 3. See also Lind, *Stap voor stap*, 115–160.
17 GFHA, Cat. no. 84, Verklaring Arthur Heinrich, 4.

At the beginning of June 1943, he and ten other pioneers were taken from Bergen op Zoom during the night to Belgium, accompanied by 'Theo', who was armed with a revolver. After quite a hike, they reached an open area where German soldiers guarded the border. 'We lay hidden between the trees and had to take it in turns to walk quickly and crouched down to cross the area. The tension was immense and one after the other we crossed the open area while 'Theo' waited with his revolver, ready to fire, until the last person reached the edge of the woods safely.'

This all went well, but when the group was several hundred meters inside Belgium, two Germans suddenly came towards them on bicycles. Everyone threw himself down on the ground. However, the soldiers, who had certainly seen them, cycled on. As 'Theo' later explained, they were afraid that he would shoot them off their bicycles.

After these wild-west scenes, which had little to do with the pacifism of the Westerweel group, a new adventure awaited them in Antwerp. The Germans hauled 'Theo' out of bed the first night after there had been a drunken argument with other smugglers. The whole pioneer group which stayed in the same hotel, had to leave in a rush because there was a chance that the guide could crack under interrogation and betray them.[18]

Willy Hirsch, who acted als leader of the group, travelled several times back and forth to Amsterdam and searched feverishly for a safe route to France. He finally came to an agreement with Julia Cole about a stay of nine pioneers in her 'inn'. Two pioneers had left on their own. The four other stranded pioneers were still there, so that there were now thirteen lodgers in *Chez Julia*. They had to stay in their room most of the time, because German soldiers also visited the inn. This went well for several weeks until the Germans became aware of their presence, possibly after an argument between Julia and a prostitute. On a Sunday afternoon at the end of June, there was a raid. However, it was not by the *SiPo*, but by the *Feldgendarmerie*, the military police. This was because of the delicate nature of the matter. *Chez Julia* was namely also a brothel for German soldiers, and the police were probably more interested in a fast solution than in a thorough investigation.

As a result of Hirsch's ballsy acting talent, the Germans accepted his well-rehearsed story that they were Dutch workers who were returning to their work in Normandy after an illegal break. As proof, Hirsch showed them several travel orders. The thirteen men were taken in a normal lorry to an army barracks, where

---

18 Asscher, *Van Mokum*, 129, 130.

they spent four anxious days. Then they were told that they could return to France after a medical examination for venereal diseases.

Because all but one or two of them were circumcised, the examination was of great importance. However, this was done by a male nurse who was in a hurry. The pioneers had to form a row together with twenty Belgians, whereby those who were uncircumcised stood between the thirteen others. Berrie Asscher, who was circumcised, stood second in the queue 'and when I heard the order "Drop your trousers", my heart started beating faster. The man asked, "Do you have any pain?" When I answered 'no', the only thing that he said was: "Go on"'. After taking this last hurdle, the group was taken to Normandy at the beginning of July, accompanied by four German soldiers.[19] It was an almost miraculously good finish to a risky adventure.

A Change of Course

Two things had become clear from the Antwerp episode. Firstly, the cooperation with 'Theo', Lelièvre's successor, had to be broken, because he did not hold to the agreements made. This also meant an end to the idea of a fast route to Spain. However, France remained a destination to flee to. The country had a German military government, so that Nazi ideology had less influence on the policy toward Jews. France was also much more thinly populated than the Netherlands, which made it easier to hide. In addition, Jews were not so noticeable in the open—especially not in the south where most people had a somewhat darker appearance.

A second conclusion from the Antwerp chapter was that taking larger groups across the border (which Shushu Simon's successor Kurt Reilinger had chosen under pressure of the increasing number of people in hiding) did not work well. Berrie Asscher's description of the journey to Belgium illustrates this. Because the transport was delayed, he had to travel several times back and forth between Gouda, Rotterdam and Bergen op Zoom. Because of the controls on the way, this was dangerous. Also, the larger transports were not well prepared. For example, according to Asscher, several passengers had much too much baggage or were too noticeably clothed. Although it was a 'warm June day', one of them, 'Leo Laub ... wore two thick coats over each other'. In combination with

---

19 Asscher, *ibid.*, 131–136.

the Jewish appearance of some pioneers, this made them immediately recognizable as refugees.[20]

Wil Westerweel was also involved with a larger transport at that time that she absolutely did not like. Kurt Reilinger had found someone ready to take the mixed group of pioneers and other Jews to France by lorry for a lot of money. The amount was paid beforehand, after which the driver would pick them up in Bergen op Zoom. Until he came, the roughly fourteen participants had to wait in the city park. 'Each of them had an orange-white striped travelling bag, which was very noticeable. A child could see that this was not just a small group of passengers ...'

The driver did not appear, but Kurt did not want to give up the escape attempt. He therefore gave the group the choice of returning to Amsterdam or checking how well the border in Putte was guarded. The fourteen people travelled there by bus, but it appeared almost impossible to cross the border. Of the four men who tried that in the night, three were arrested. The rest took the bus back to Roosendaal, but got no further because of the curfew. Wil 'broke out in a cold sweat' every time that she saw the group. However, she stayed with Kurt because she did not want to abandon him 'in this terrible situation'. She went to a hotel with the women on a wing and a prayer, hoping that the owner would not betray them. Kurt did the same with the men. 'Miraculously, we were both lucky', Wil recalled after the war.[21]

There is no complete picture of the escape attempts to Belgium and France. Some pioneers left on their own for France at the end of 1942. Besides the previously-mentioned Schraga Engel of the Deventer Union and Rolf Schloss from Gouda, this group included, for example, Willi Simons from Loosdrecht. He succeeded in reaching Switzerland, but later returned to France to fight in the *Maquis*, where, he went missing in action.[22]

It is also unclear about what happened to the larger groups that left for France via Belgium in the period of June up to and including August 1943. There were probably four or five such groups. The impression is that Kurt Reilinger cooperated in these transports with others, Jews and non-Jews, who had papers and/or money. Non-Jews, mostly students were also involved in several transports, to make the group less noticeable.

---

20 Asscher, *ibid.*, 127, 128. Laub was arrested later in Bergen op Zoom, when he went for a walk, against the express order of 'Theo'. He was sent to Westerbork as a 'punishment case' and from there to the east, where he died.
21 Westerweel, *Lijn of cirkel*, 71, 72.
22 Pinkhof and Brasz, *De Jeugdalijah*, 109.

Sometimes this went well, but more often than not it seemed to go wrong. In June 1943, Max Krzeszower, an 18-year-old from Loosdrecht, was part of a group of 'largely Dutch Jews' who were betrayed and fell into German hands at the Belgian border. An officer told them that he knew that they were Jews with false papers. Everyone who does not report voluntarily before this is proved in a control in The Hague, 'will be summarily executed tomorrow morning at this wall behind you', he menaced.

Max, who knew that he had good papers 'that were not on the black list', was the only person in the group who did not report. Despite intimidating interrogation, he insisted that he was a Dutchman with an Indonesian background, who was on his way to France to go to work there. He was finally allowed to sign for the *Arbeitseinsatz*, from which he quickly escaped to go to France, with Joop Westerweel's help.[23]

A New Network in France

At the beginning of August 1943, Kurt Reilinger went to France to restore Shushu Simon's old contacts with the Jewish resistance and to lead the growing group of pioneers staying in France. After this, the transport across the border was mainly carried out in small groups of two or three people. Later, several larger groups would also go across the border with false travel documents. Besides Joop Westerweel, Menachem Pinkhof, Jan Smit, Bouke Koning and Tinus Schabbing also began to be involved with the transport to France. Tinus Schabbing was a new member of the group.

Schabbing, born in Amsterdam in 1918, had taken a German-Jewish married couple, acquaintances of his wife, to Antwerp in the spring of 1943 as the first part of their route to Switzerland. After a difficult youth, Schabbing had become a metal worker, but to avoid the *Arbeitseinsatz* he was in an agricultural work project at that moment. He had been active in the socialist youth movement AJC, but was mainly motivated by the injustice done to the Jews. His journey to Antwerp somehow became known in *Hechalutz* circles and several weeks later, in June or July 1943, Lore Durlacher visited him and asked: 'Tinus, would you go for us? That's how it started, quite simply really. Well, you'll hear from us if someone has to go away'.[24]

---

23 Frans van der Straaten, *Om nooit te vergeten* (Mijnsheerenland: self-published), 89, 90.
24 GFHA, Cat. no. 204, Interview Yigal Benjamin-Tinus Schabbing, 22 januari 1990, 1–7. In another statement from 1960, Schabbing reported taking the pioneers to Brussels. The destination possibly changed. See also Interview Sytske De Jong-Schabbing, 13 oktober 1999.

Joop Westerweel also visited him. He 'wanted to take his measure of me', thought Schabbing. Joop apparently liked his measure because, from the summer of 1943, Tinus Schabbing kept busy taking pioneers in groups of two or three to Belgium. He was not at all pleased by the Budel route, which he visited with Joop: 'a fairly long straight road ... all open and bare, that didn't sit well with me at all'. Instead of Budel, he found his own route via Putte. He made friends with a café owner who kept his false Belgian identification card for him. A farmer allowed him to walk over his land towards Belgium. He then took the pioneers across the border via 'all possible rear gardens', where they came out close to a tram stop to Antwerp.

Tinus Schabbing worked in roughly the same manner as Chiel Salomé. Lore Durlacher told him where he had to pick people up, but not who they were. He then briefly told them what they had to do, i.e. travel to Putte by train and bus, but independently of each other. They got out in Putte and the two of them had to follow him at a distance. So that he could always be easily spotted, he wore a red scarf. The last stage of the journey was the tram to Antwerp, where pioneer Max Windmüller (cover name 'Cor') took charge of them.

Tinus Schabbing made about twenty journeys to Antwerp. Once, out of curiosity, he went to Paris with 'Cor'. In addition, he cared for people in hiding, and later helped on one or two occasions with an escape from Westerbork. Tinus had papers from the *Gemachtigde voor de Arbeid* (Authorized Representative for labor) to travel through the country as an agricultural inspector. He was arrested twice when crossing the border. The first time, he was able to use these papers. The second time, when Frans Gerritsen and Jan Smit were with him, they said they were tobacco smugglers. This was accepted after the usual interrogation, but they lost their tobacco and a large amount of their money. Because of his limited income, Tinus later received a monthly payment of 200 guilders from the *Hechalutz*.[25]

In France and Belgium, a new pioneer network arose led by Kurt Reilinger. Max Windmüller and Willy Hirsch were key figures in the transport. Windmüller was a son of a cattle dealer and butcher born in the eastern Friesian town of Emden in 1920. When his father's work permit was withdrawn after Hitler's rise to power in 1933, the family went to the Netherlands. They settled in Groningen, where Max was active in the Zionist youth movement. At the end of the 1930s, he took part in the Deventer *hachshara* on a farm in the surroundings of Assen. In 1939, he was about to sail for Palestine together with his older broth-

---

25 Ibid.

er Isaak on the *Dora*, when Ru Cohen called on him to remain active in the Netherlands.

Max returned to Assen and worked there until he went into hiding in The Hague at the end of 1942. In an attempt to flee to Belgium, he was arrested in August 1943 and taken to Westerbork. From here, he escaped several days later in a car with the laundry. With papers made by Frans Gerritsen, he then went to Belgium, to act as a link with the transport of pioneers to France. Later he worked from Paris. Max Windmüller, who was known in the resistance as Cor, had an even and good-humored character. During the train journeys, he often played all sorts of songs on his harmonica. These included French request numbers, and also 'Jewish' songs such as the well-known *Hava nagila* and various pioneers' tunes. When enthusiastic fellow passengers asked what these were, Max said that they were Dutch folk songs.[26]

In Antwerp, Ernst 'Willy' Hirsch had turned out to be a natural talent for organization with nerves of steel. The member of *Hashomer Hatzair* was born in 1916 in Aachen, just across the border in Germany, and spoke fairly well Dutch. In 1939, he went to the Netherlands, where he worked in the *Werkdorp* and later at individual *hachsharas*. Willy Hirsch was a simple, practical man, who had courage and perseverance. All together, this was an excellent combination for operating in a war situation. Hirsch went to France to arrange the contacts between the pioneers and also functioned as contact with the Jewish resistance in France.[27] Later on, Reilinger's network expanded even further.

New Papers

Initially, pioneers went to Belgium and France across the 'green border', i.e. over forest paths and dirt roads to avoid the official border posts. The Dutch-Belgian border was an especially dangerous obstacle. Once it was passed, people could breathe more freely—and once France was reached, 'the relief was even greater', according to Menachem Pinkhof.[28] Gradually, pioneers were able to travel using the false travel documents that became available instead of having no choice but to traverse the always risky 'green border'. The smuggling route was far from superfluous, however. The danger remained that the false papers would be recognized as such or were not valid. The first time Shushu Simon crossed the border,

---

[26] Klaus Meyer-Dettum, *Max Windmüller 1920–1945* (Emden: Arbeidskreis Juden in Emden, 1997), 1–32. See also Siegel, *Locomotieven*, 148–149.
[27] GFHA, Cat. no. 92, Informatie over Ernst Hirsch, probably written by Kurt Benjamin.
[28] GFHA, Cat. no. 173, Interview Haim Avni-Menachem Pinkhof, juli 1961, 4.

Fig. 20: Max (Cor) Windmüller (right), here with fellow pioneer Harald Simon, accompanied pioneers from Brussels through France. Max died shortly before the liberation on one of the death marches from concentration camps. Harald Simon, who was arrested in Rotterdam in 1943, was murdered in Auschwitz.

he had a false French work pass, which was worth nothing in Belgium and little in France. On one of his journeys to France, Simon took blank travel forms for foreign workers with him. As Willy Hirsch had shown, these could be used in Belgium and France.

These travel forms were not valid, however, to pass the Dutch-Belgian border. This required a document with a watermark that could not be forged and that was issued by the *Prüfstelle* (inspection authority), which was established on the Champs Elysees in Paris.[29] This office assessed all travel documents of German military and government institutions in West Europe. The inspection authority for Eastern Europe was established in Kiev.

The first person who had a document to cross the border was Heinz Fränkel, who was born in Vienna in 1921. Eighteen years old he fled to the Netherlands in 1939, working at individual *hachsharas* with farmers in Gelderland and Overijssel. At the beginning of 1943, he went into hiding in a hotel with a false identification card, was betrayed, but said he was a half-Jew; they therefore allowed him to do all sorts of jobs in the prison of Enschede. Finally, he fled to Amsterdam, from where he went to France in the spring or summer of 1943, with the aid of false papers from Kurt Reilinger. In France, he found work near Dieppe.

With his blond hair, Fränkel looked 'very un-Jewish' and spoke German with a Viennese accent, which he explained to his German director by saying that his mother came from Vienna. This created a bond between them. After some time, the director asked if he wanted to recruit workers for his company in the Netherlands for a premium of 1,000 French francs. Because people were continually moving on, they always needed new workers. Fränkel agreed, after which they went to Paris together, where the director's brother was a liaison officer for the German air force in Belgium and northern France. Although he did not have the authorization, he wrote a letter with which Fränkel received a pass for the French-Belgian and Belgian-Dutch border plus 50 blank leave passes. His official position was now *'Transportführer für Wehrmacht-Zivilarbeiter'* (Head of transport for civilian workers for the German army).[30]

Heinz Fränkel, who had a great deal of nerve, travelled back to Amsterdam and reported for lodgings at the German army home, where he was given a room with a telephone. He then took the papers to the *Hechalutz*, where Kurt Reilinger said that he found their use much too dangerous. Hans Ehrlich, another colleague at the Hechalutz, did not agree with him and offered to test the leave passes personally at the *Ortskommandantur* (local command post) on the Museumplein. With a false identification card (the building was forbidden for Jews), he got approval for Fränkel's leave passes. Using these passes, a first transport of eleven

---

29 Ibid., 3–8. The method of making false documents look 'genuine' to the inspection authority was also followed by Jaap Penraat's group, whereby the photographer Krijn Taconis was involved. They worked together with the French resistance and took about 400 Dutch Jews to Spain. See also Hudson Talbott, *Forging Freedom* (New York: G.J. Putnam's Sons, 2000). TV Program VPRO, Spoor Terug, Interview Jaap Penraat dl. 1, 17 juni 2001.
30 GFHA, Cat. no. 56, Verklaring Heinz Fränkel, probably 1956, 1–5. Fränkel, who was an electrical engineer, went to the United States after the war. He changed his name to Henry Frank.

'*Atlantic Wall* workers', six pioneers and five Dutch students, set forth in November 1943. However, in Roosendaal it appeared that the requisite *Prüfstelle* document was missing. After a hefty argument between Fränkel and the military police, whereby Ehrlich mediated and calmed the *Transportführer,* the group was allowed through.[31]

Because it was clear that this would not work a second time, Menachem Pinkhof went several times from the Netherlands to the *Prüfstelle* to get papers for pioneers in the late autumn of 1943. Pinkhof travelled with the small folded false identity papers in the double bottom of a butter dish to Paris, where a sort of forger's center had been realized under Reilinger's leadership. Here, authorized employers' statements were added with false stamps. With these papers, Pinkhof (who had a false identification card on which his profession was given as 'military engineer'), went to the Champs Elysées office, where he received the coveted watermark document. Ehrlich travelled from Normandy to the *Prüfstelle* several times to get leave passes approved. The great risk here for both Pinkhof and Ehrlich was a possible check with an employer stated on the papers, who knew nothing of it. However, this never happened.[32]

In consultation with Kurt Reilinger and Menachem Pinkhof, Heinz Fränkel made several trips accompanying groups of ten to twenty 'workers' to the French coast. They sometimes even used a German army train that went from Germany via Maastricht and Brussels to Lille once a week. Approximately six pioneers took part in each group. In a post-war statement, Fränkel estimated that he had taken at least 150 people to France. However, this number appears too high. According to Pinkhof, there was a total of four transports, whereby the last one went wrong. The difference between estimates might be explained because Fränkel also took groups without pioneers. It is likely that several dozen Palestine pioneers in total went to France with the false leave passes.[33]

A last transport at the beginning 1944 failed. This was because a very deserving Dutch resistance woman had convinced Pinkhof and Fränkel to take her German boyfriend (a non-commissioned officer who had deserted and was being sought) with them to France. He was recognized during a train control, and all but one of the roughly twelve-man group were arrested. The deserter was executed a short time later. When this also threatened to happen with *'Transportführer'* Fränkel, he admitted that he was Jewish. The Germans sent him to

---

31 GFHA, Cat. no. 48, Verklaring Hans Ehrlich (born 1919 in Dresden), probably 1956 or 1957. Supplementary correspondence of Hans Ehrlich with Kurt Benjamin, 23 maart 1990.
32 GFHA, Cat. no. 173, Interview Haim Avni-Pinkhof, 6, 7.
33 Ibid. See also GFHA, Cat. no. 193, Verklaring Walter Rosenberg, 26 april 1956, 1–3.

the punishment barracks of Westerbork. Menachem Pinkhof, who had also been arrested, was given permission to take him to the Bergen-Belsen concentration camp, where the circumstances (living conditions) were more favorable. Heinz Fränkel survived the war.[34]

As a result of the arrests at the beginning of 1944, the Westerweel group called a halt to this method of group transport, and reverted completely to using the old method of the 'green border.'

In a year's time, from spring 1943 until spring 1944, about 150 Palestine pioneers left the Netherlands in various ways to go to France.

Working for the *Organisation Todt*

Travelling to France along all sorts of secret routes is one thing, staying out of sight of the Germans is another. This meant that the Palestine pioneers had to have a cover story that justified their stay in France. This cover story was provided by the previously mentioned *Organisation Todt* (*OT*), the company commissioned to build the *Atlantic Wall*. The idea of using pioneers arose in the Westerweel group at the same time (the spring of 1943) as the idea of hiding in the German *Arbeitseinsatz*—and was possibly also inspired by it. The requisite false identification cards were available, but papers to pass the Belgian and French border were missing. 'Theo', Henri Lelièvre's successor as guide to France had a solution. Dressed in *OT* uniform, he took four pioneers to Belgium around 20 May 1943. They acted as Dutch workers who had stayed at home instead of returning to their work in France, and were now being taken back.

When these four pioneers, led by the slightly older Emil Glücker arrived in Paris, a city where they knew nobody, they happened to hear from Dutch workers that there were jobs in Bordeaux. When they arrived there, they met a Dutch contractor at the station who worked on the docks in Bordeaux and La Rochelle for the German company *AEG*. In La Rochelle, the four pioneers were set to work on a submarine base. They passed along their experiences to their contacts in the Netherlands.[35] This is how working in France became an alternative to the fast route to Spain, which had proved to be impossible.

In Germany, the pioneers employed there had to work in a strictly controlled industrial structure. After several months, this led to their exposure and arrest.

---

34 Ibid., Cat. no. 56, Verklaring Fränkel, 3–5.
35 GFHA, Cat. no. 203, Verklaring Erich Sanders, 3 april 1956. The statements of Emil Glücker and Alfred Dubowski largely correspond with those of Erich Sanders.

The work on the German defense works in France had, however, quite a different, chaotic character. After Pearl Harbor and the German declaration of war on the United States in December 1941, Hitler had ordered the building of the *Atlantic Wall*. Later, after the Allied raid on Dieppe in August 1942, the building plans were considerably expanded. Along the coast of France, Belgium and the Netherlands, more than 15,000 concrete reinforcements had to be built so that the German army would be able to immediately repulse any possible invasion. The German military engineers and the *OT* were given the order to build the wall.

The *OT*, founded in 1933 by the mechanical engineer Fritz Todt, initially kept to the construction of German motorways. From 1938, the *OT* constructed defense works such as the *West Wall*, which ran from the Netherlands to the Swiss border. Now the building of the *Atlantic Wall* was added to the work. On paper, the *OT* formed an impressive whole, but in practice, the organization struggled continually with shortages of workers and materials.[36]

The shortage of workers for building this wall provided the opening needed by the Palestine pioneers; an understanding of how the *OT* coped with these shortages helps in explaining the pioneers' activities in France. The thirteen 'Antwerp' pioneers supplied by Willy Hirsch arrived at their place of work in Dannes-Camiers near Boulogne in July 1943. Since there was nobody to check if they had worked there earlier, as they claimed, they were deployed immediately into working parties under German supervision to pour concrete in the construction of a bunker. This was work that they did not know but quickly learned.

Even when pioneers were recognized as Jews, this apparently did not interest the leaders of the company where they worked. Arthur Heinrich from Loosdrecht describes how a leader in Dannes-Camiers 'had immediately accused us of being Jews and wanted to check our papers, but he did not pursue the matter any further'. Pioneers at other work places had similar experiences. Despite their denial, they were denoted as being Jews, but nothing happened.[37]

Many pioneers did not like having to work for the Germans. For this reason and also with a view to their safety because of the dangers, Willy Hirsch quickly decided to reconnoiter a route to Bordeaux in southern France, which was close

---

36 Rémy Desquesnes, "L'Organisation Todt en France (1940–1944)," *Histoire, économie et société*, 1992, no. 3, 535–550. In France the OT had in all 291,000 employees, of which 15,000 were German *OT* executives. 110,000 of the workers were French and from North-Africa. Of the 165,000 foreign workers 50,000 were *Zwangsarbeiter* from Eastern Europe, 35.000 came from Poland and Czechia, 15.000 were volunteers from Spain and 20.000 came form Belgium and the Netherlands.

37 GFHA, Cat. no. 84, Verklaring Ascher Heinrich, 5. See also Cat. no. 43, Verklaring Alfred Dubowski, 2.

to Spain. His disappearance was not a problem. Berrie Asscher, one of the thirteen pioneers, had already noticed that there was not much left of the German thoroughness in the organization, and that 'it was a coming and going of workers' at *OT*. After just over three weeks in Normandy, Berrie and two comrades also travelled to Bordeaux via Paris.[38] Others from the group remained working for some months in Dannes-Camiers.

The Stream Starts to Move

Emil Glücker's four pioneers and the thirteen pioneers supplied by Willy Hirsch from Antwerp were the first Palestine pioneers who went to work at the *OT* in France. The *chalutzim* who came after them had essentially the same experiences. The 22 year-old Walter Rosenberg, having fled from Dortmund to the Netherlands in December 1938, found work in the *Wieringermeer*. In the autumn of 1943 he travelled by German army train from Maastricht to Lille with a mixed group of approximately 20 pioneers and students. Their travel guide was Heinz Fränkel. Near Lille, they found work with a Dutch company that worked for the *OT*.

After a fatal workplace accident, the pioneers fled in panic because they feared that an investigation into the accident would show that their papers were false. Walter just set out for Paris, where he reported to the *Rüstungsinspektion (armament inspectorate)*. They sent him to a company that installed sanitary equipment and heating for the German army. Because his work was not good enough, he was quickly fired. He then worked as an unskilled laborer in different factories. Through the French labor exchange, he got a job at *Henschel Werke*, which made aircraft engines near Paris. The last months before the liberation, Walter sorted letters for the German military mail service.[39]

According to his own statement, Walter Rosenberg stayed in France in connection with the *Hechalutz*, but was forced by circumstances to go his own way. Other pioneers kept closer contact with each other and the leaders. In Auffay, a small town in Normandy between Dieppe and Rouen, there was for some time a sort of shelter for pioneers who had arrived from the Netherlands. This was run by Max Bonn and his wife Rie, who had arrived in France together with two other pioneers in October 1943.

Max, a 24-year-old from Amsterdam, had a rather unusual background for a Palestine pioneer. Until June 1940, he had attended a school for professional

---

38 Asscher, *Van Mokum*, 140, 141.
39 GFHA, Cat. no. 193, Verklaring Walter Rosenberg, 1–3, 26 april 1956.

non-commissioned officers in Haarlem. After the school closed, he prepared himself for the *hachshara* with the Deventer Union, and began his training in June 1941—taking his pistol from officers training with him. Later, he gave the weapon to Kurt Reilinger. In the autumn of 1942, Max was called for a medical examination for a work camp for Jews in the Netherlands. He went into hiding, but was arrested by the police fourteen days later.

Max Bonn escaped shortly afterwards, after which Ru Cohen (who had strongly advised him against trying to flee the country) helped him get a job, together with his wife, at *Het Apeldoornse Bos*. When the Germans cleared out the psychiatric institution at the beginning of 1943, they fled to Soestduinen, where they went to live in a guesthouse. Rie Bonn, who did not look Jewish, was active in the Dutch resistance as a caretaker of Jewish children in hiding. After several months, she was betrayed and locked up in the Hollandse Schouwburg. It was, however, possible to buy her release for 500 guilders. In the autumn of 1943, Kurt Reilinger informed Max and Rie about the possibility of going to France.

In mid-October 1943, Joop Westerweel took Rie and Max across the Belgian border. Accompanied by Willy Hirsch, they then went with six other pioneers to Paris via Antwerp and Brussels. From Paris, they were sent to Auffay, where they went to work for the Belgian company De Moll. They were quickly able to rent several rooms from the local baker, where they and later also other pioneers lived.

Max first worked as a carpenter and later in the office of De Moll, where he could take travel and residents permits with him. Rie did the household for the pioneers. Various pioneers stayed there for longer or shorter periods, including Paul Landauer, Hans Flörsheim, Emil Windmüller, Rolf Rothmann, Emanuel Benjamin and Lore Sieskind (who assisted Rie in keeping the household running). Several of them had previously worked at the Paris airfield Le Bourget, but the Germans there were sending all Dutch workers to the *Arbeitseinsatz*. Obviously, the pioneers were not keen on this. Through the help of the mayor, they were able to rent a house in Auffay.

When it looked like the malpractices at De Moll were going to be exposed, Max and Rie Bonn left for Dieppe, where they worked for a Dutch company. The documents that Max was able to steal at that company were mainly intended for Dutch students who also worked there for the *OT*. During an arrest, however, one of them cracked, which led to Max's arrest in spring 1944. After the Allied invasion, he was put on transport to Germany as a Dutch worker, but he escaped while still in France. He then went to Paris, where Rie was in hiding, and waited for the liberation.[40]

Southern France as the Next Step

The first groups of Palestine pioneers who arrived in France worked mainly in the north of France. After Willy Hirsch had established that it was quite easy to get work with the *OT* near Bordeaux, many of them went to the south. After Kurt Reilinger arrived in France, several pioneers also went to Toulouse. Especially those who arrived in France after January 1944 went there directly via Paris.

Through their work at *OT* and the connections that they built up there, the pioneers quickly learned their way in France. When Berrie Asscher arrived in Bordeaux, he immediately went to 'Albert the Belgian', a contractor for building companies in the south of France. Albert sent Berrie to the Belgian company Van Craenenbroek, which built barracks for the German army in Gradignan, near Bordeaux. During this period, civil work was almost impossible to find. Willy Hirsch and several other pioneers also worked at this company.

At Van Craenenbroek, the pioneers received a travel pass issued by the Vichy government, which also gave them the right to ration cards for food and textile. The pass gave Berrie Asscher 'a deep sense of security'. After several weeks, he left for the Ile d'Oléron to work for a German company laying underground cables. Other pioneers worked near Bordeaux building first roads and later an ammunition depot in Jonzac.[41]

Female pioneers such as Paula Kaufmann, who had arrived in France in the beginning of 1944, were given office jobs with German companies—partly because of their knowledge of the language. Pioneers who did not want to work for the *OT* sometimes succeeded in getting work with French farmers. This happened, for example, to Lore Sieskind, who worked at a farm near Toulouse. The 22-year-old Ernst Asscher, who had worked in the *Wieringermeer* and who had been taken across the border by Tinus Schabbing in November 1943, also found work with a farmer, who was actually a son of the Van den Akker family, where he had hidden in Amsterdam. The farm was located in northern France,

---

[40] GFHA, Cat. no. 18, Verklaring Max Bonn, 1–4, december 1956. See also Flörsheim, *Über die Pyrenäen*, 80–89.
[41] Asscher, *Van Mokum*, 143–153.

where another pioneer already worked. Later, Paula Kaufmann took Asscher to Paris.[42]

The number of pioneers in the Bordeaux area had so increased at the beginning of 1944 that Joop Westerweel went to Sevenum to ask the French-speaking Eugénie Boutet if she was willing to lead a new *Beth Chaloets* in La Bouheyre, near Bordeaux, as many pioneers worked in forestry in this small place. Boutet declined, however. At the beginning of April 1944, Kurt Reilinger put the same question to Lilo Spiegel. Up to then, she had been taking care of the wounded Norbert Klein in Haarlem, but she agreed to the proposal. Much to her annoyance, however, nothing came of the *Beth Chaloets*.[43]

The circumstances under which the Palestine pioneers worked in France were good in comparison to those in the Netherlands. The food situation was much better and reasonable salaries were paid. With the Vichy pass or other genuine or false papers, it was also possible to visit cinemas for German soldiers and army canteens. Not all pioneers did this, but those who did played the role of Dutch workers, so that they were not noticed.

Much more important than these material matters was the feeling of freedom that the pioneers in France experienced. Berrie Asscher described this as follows. 'After only two months of having left the Netherlands with all its limitations, raids, 'stars' and persecutions, it was such a strange experience for us to be able to move freely as Dutch workers between the enemy that we could hardly believe it ourselves at times. The transition was too fast and the change too great.'

There were certainly pioneers who, as a result of the sudden freedom, forgot that dangers still existed and who became overconfident. For example, several of them travelled with admittedly valid leave passes to the Netherlands for a short holiday. However, they used false identification cards, so that they still ran great risks. Berrie Asscher also went on leave to the Netherlands and wondered later if he had not 'gone completely off his rocker' in going 'to the scary tiny Netherlands with all its dangers', instead of staying in the relatively safe south of France. He mainly ascribed his 'completely irresponsible' behavior to his youthful age.[44] In combination with the feeling of relative invulnerability arising from the sudden freedom they experienced, this will also have been an important factor with the other people on leave. Although most pioneers made it through this

---

42 GFHA, Cat. no. 65, Verklaring Paula Kaufmann en Ernst Asscher, februari 1957, 1–5.
43 GFHA, Cat. no. 400, Interview Mirjam Pinkhof-Eugénie Boutet, juni 1991, 1–5. See also GFHA, Cat. no. 230, Verklaring Lilo Spiegel, 1–4, 1955. GFHA, Cat. no. 217, Verklaring Lore Sieskind, 7.
44 Asscher, *Van Mokum*, 155, 156.

time, one pioneer was arrested during leave and sent to Auschwitz; he did survive, however.

* * *

With hiding places opening up in northern Limburg and especially a route found—albeit through trial and error—to France, the year 1943 (which had started unfavorably with the arrest of Shushu Simon and Rik Lelièvre) could still be relatively favorably concluded. An important aspect here was the fact that the Westerweel group was successful in attracting new members, such as Chiel Salomé and Tinus Schabbing. Together with Max Windmüller and Willy Hirsch, Kurt Reilinger built up a network in France to house pioneers from the Netherlands. Menachem Pinkhof and Heinz Fränkel provided good travel documents, at great risk to themselves.

The work for the *Organisation Todt* on the *Atlantic Wall* offered a good cover for the stay of the Palestine pioneers in France. As a result of the less-rigid occupation regime in that country, there were clearly fewer dangers for Jews in hiding. The disadvantage was that the *Atlantic Wall* was a military project. There were, however, many ways that could be found to slow down the work and even carefully commit sabotage. Later, the pioneers discovered that it was possible to do non-military work on a limited scale.

Setbacks included the failure of attempts to hide in the *Arbeitseinsatz*, as well as the German capture of several larger transports of refugees to France. Although there is no clear picture about this, almost certainly dozens of refugees, including several Palestine pioneers, were arrested. The main cause of these fiascos can be found in the great pressure under which the Westerweel group had to work because of the rising number of pioneers and other Jews who sought a hiding place or wanted to flee from the Netherlands. As this pressure increased, so also did the risk of betrayal.

9 Westerbork and Beyond

Camp Westerbork is mentioned several times in previous chapters as a place where pioneers and other Jews were taken for further transport. This camp was founded at the beginning of 1939 to house Jewish refugees from Germany. The first inhabitants arrived in October of that year. The remote camp consisted of wooden barracks and lay in a thinly-populated area in the province of Drenthe. The surroundings consisted mainly of heaths and dark-brown marshes. 'For as far as the eye could see, there was nothing more than the sparse, desolate ground with here and there a shabby farm', noted a detained pioneer.[1]

In July 1942, the legal status of Westerbork changed. Under the name of *Polizeiliches Durchgangslager* (Police transit camp), it then became an institution led by the *SiPo/SD*, with a barbed wire fence, watchtowers and a moat. Dozens of *SS* men were present in the camp, but they entrusted most of the management, including the guarding of prisoners, to Dutch civil servants and Dutch marechaussees (MP). The Constabulary behaved itself with regard to the camp inmates mostly quite well. The original, mainly German Jewish inhabitants were given positions in the management and organization of the camp. This gave them a privileged status and a place on the 2,000 names of the so-called *Stammliste* (master list) of camp prominents, who in principle were exempt from deportation to the east.[2]

As the name 'transition camp' indicates, most of the more than 100,000 Jews who entered Westerbork from July 1942 stayed only temporarily. Initially, a train comprised of closed freight wagons containing between 800 and 2,000 people left twice a week for Poland. In all 107,000 Jews were deported of whom about 5,000 survived. From February 1943, the train left once a week. The number of Jews to be transported by train was determined in Berlin. In Westerbork, the camp administration (the *Registratur*, in the strongly German-tinted camp idiom) then determined who had to go to the east. The administration, where mainly so-called *alte Kampfinsassen* (old camp inmates) worked, was supervised by the *SS* staff, under the command of *SiPo*/SD officer Albert Gemmeker.

The evening before the transport, the names of the people involved would be read out in the barracks. This caused heartbreaking scenes. Despite the unpleasant circumstances and the unreal situation in which they found themselves,

---

1 Kochba and Klimov, Manuscript, hfdst. 6, 238.
2 Guido Abuys and Dirk Mulder et al., *Een gat in het prikkeldraad* (Hooghalen/Assen: Herinneringscentrum Westerbork, 2003), 7, 8.

many Jews regarded a stay in Westerbork as a last link with normal existence in the Netherlands. The train of the next day put an end to this illusion.

Palestine Pioneers in the Camp

The first group of more than 50 Palestine pioneers, including approximately 20 juveniles, arrived in Westerbork in July 1942, with the first Amsterdam transports. They were almost immediately sent through to Auschwitz.

The second group included the 36 pioneers with a Palestine certificate from Elden, in Gelderland, who arrived at the beginning of October of 1942. They escaped immediate transport to the east through the protection of Manfred Samson's father. The Samson family, with eastern Jewish roots, came from Leipzig. Samson senior, who had Polish nationality, was arrested at the beginning of 1938 and was then expelled to the Netherlands, where his brother lived. Initially, he stayed in a home for refugees in Amsterdam; later, he had to go to Westerbork. In 1939, the rest of his family joined him there.

Manfred, who was born in 1923, was arrested during the *Kristallnacht* and then also expelled from Germany. He went first to the Gouda *hachshara* and then, as holder of a Palestine certificate, to Elden. His father belonged to the first group of inhabitants of Westerbork, which gave him a privileged position and great influence with the camp leaders. Samson senior was the head of the office that issued passes for journeys outside the camp. Later, he was head of the so-called *Arbeidsbureau* (employment office).[3] Through his advocacy, the Elden pioneers were allowed to stay together as a group. An important argument here was that they were used to hard work and could also do this in Westerbork.

From the end of 1942 and the beginning 1943, the Palestine certificate (an entry visa), also offered the people from Elden a certain protection. The Jewish Council then made a list, divided into different categories, with the names of 1,500 holders of such documents. These individuals would become eligible for exchange with Germans who were staying in Palestine. In July 1943, the so-called Palestine list was largely declared invalid. After this, it once again grew to almost 1,300 people at the end of the year.[4]

To be able to stay in Westerbork for a longer time, it was of great importance to be placed on one of the different lists. Such a place provided a so-called *Sperr*,

---

[3] Schlomo Samson, *Zwischen Finsternis und Licht* (Jerusalem: Rubin Mass GmbH, 1995), 1–166.
[4] Presser, *Ondergang*, dl. II, 43, 45. For a complete summary of all categories that were eligible for a place on the Palestine list, see Wasserstein, *Gertrude*, 158–164.

which kept the holder safe from deportation for the time being. However, just as this was the case with the Palestine list, these lists were regularly completely or partially declared invalid by the Germans. But being placed on a list, however, often won some time, with regard to deferring deportation.

Whoever was either not on a list (which included the majority of the Jews who were brought in), had no money or could not call on what was called 'Vitamin R' (i.e. relations) in the camp, mostly stayed only a short time, one or two weeks, in Westerbork. Their stay in the camp ended with a train ride to the east.

The original Elden group was supplemented by pioneers who were already in the camp or came in later, mostly as detainees and punishment cases. Paul Siegel of the Deventer Union also arrived with this status in Westerbork at the end of November 1942. As described, he was taken with three other pioneers from their hiding place by the Dutch police, thrown into the Arnhem prison and then transported to the camp. At Westerbork, they were locked up in the punishment barracks. Placement in these barracks almost always meant that the person involved was slated for transport to Poland on the next train.

Manfred Samson, who could move about freely in the camp and who acted as a contact person for the pioneers, told Paul that there would be a train the next day. That night, in the punishment barracks, he experienced the calling of the names of the people who had to go by train. With 'baited breath', the pioneers listened to the barracks elder, who read the names in alphabetic order. 'I sat there white as a sheet, my teeth chattering and shaking with fear. I hardly heard the names, but waited for the letter S to be read ... Then it started. "Sanders, Segel, Silberstein". I shrunk back. Was that my name? Did I hear it well? That was the last name that I wanted to hear.'

However, Paul's friends reassured him. None of their names was called. Wet from a cold sweat, Paul looked around him and 'saw people running back and forth, quietly sobbing and screaming nightmarishly. Everyone who had heard his name called knew that he would leave the Netherlands in several hours' time— perhaps forever'.

That night made a deep impression on Paul Siegel, who found out only months later how he had been saved. Mau Reichenberger, the head of the Jewish Council in Gelderland, who was an acquaintance of Siegel, had asked the Germans if Siegel could be released. This was not possible. However, as a favor to Reichenberger, the arrested pioneers were allowed to leave the punishment barracks and join the Elden group. Just as the exemption from transport of this group, this was a striking example of the operation of the extremely important Vitamin R in the camp.

When Siegel, his three comrades and a friend (the non-Zionist Herman Italiaander, who they had persuaded to join them) arrived in their new barracks 64, there were approximately 70 pioneers already there. The core was formed by the group from Elden. Furthermore, there were *chalutzim* from the Deventer group and from the religious *hachshara* training courses in Enschede and Laag-Keppel, as well as several people from Gouda and Loosdrecht. The number of pioneers would rise until mid-1943 to about 120. Later, several dozen mostly young Zionists also came, so that in September 1943 the *hachshara* group consisted of approximately 150 people.[5]

Fig. 21: The hachshara group, guarded by several MP's at work in the field outside the Westerbork camp

The Palestine pioneers had, for various reasons, an unusual status in Westerbork. They formed a unit with a strong moral code, in which Zionism played an important role. The Amsterdam journalist Philip Mechanicus, a sharp observer who was imprisoned in the camp, noted in his diary that he regarded this Zionism as a positive development for European Judaism. In his opinion, it could form an alternative for both the rigid orthodoxy and the unprincipled materialism that he thought he saw in many people. In Westerbork, he regarded the

---

[5] Siegel, *Locomotieven*, 108, 109. According to other sources, the *hachshara* group never had more than 120 members.

members of the *hachshara group* as 'the predecessors of the new generation'. They were often well educated and 'filled with a strong spiritual awareness, physically well built, strong and courageous'.⁶

According to both Mechanicus and Paul Siegel, Commander Gemmeker was also impressed by their attitude. He therefore reserved certain tasks, such as the care for his garden and house, for the pioneers.⁷ It is essential to bear in mind, however, that the Germans were mainly interested in organizing the transport to the east with as few problems as possible. The tasks given to the *hachshara group* were part of this façade.

Unknown Destination

From the beginning of the deportations, there was much speculation in the Netherlands about these circumstances 'in the east'. This speculation took place mainly in Jewish circles, of course, but also outside. After the first transports, positive messages from the deportees arrived. According to these messages, the men would have to work in factories or for farmers, and the women would have to work in the camp. The messages, which were probably dictated by the Germans, were not believed by everyone.

People in Westerbork tried to get more certainty about the camps from messages that the deportees could send. This appeared possible when it was clear that the same train was used for the transport to Auschwitz each time. Several volunteers were asked to write their experiences down and to hide the notes in an air duct of a wagon. In Westerbork, cleaners would then remove the notes. This did not supply any usable information about Auschwitz, because the people involved got off the train without having yet seen the circumstances there. 'Mostly, the notes described the train journey, with the emphasis on the circumstances, the landscape and the stations.'⁸

When the Palestine pioneers took over the cleaning of the trains, they arranged with two pioneers who were to be transported that they would stick notes with information about the journey at a previously-agreed place in a wagon. But this did not provide any new information. Inspired by Frans Gerritsen, the possibility of sending several people to Poland to investigate the situation themselves was considered. However, nothing came of this. In the *chalut-*

---

6 Philip Mechanicus, *In Dépôt, dagboek uit Westerbork* (Amsterdam: Van Gennep, 1964), 16, 17.
7 Mechanicus, *ibid.*, 82. See also Siegel, *Locomotieven*, 95.
8 Abuys and Mulder, *Een gat*, 24, 25.

zim, many people were aware of *BBC* radio transmissions that reported the mass murder of Jews in Poland in autumn 1942.[9]

At the beginning of 1943, a confirmation of these murders arrived through a message from Frits Lustig, a German pioneer of the Deventer Union. After his deportation, he was put to work in the Buna-Monowitz camp, which belonged to the Auschwitz complex, where synthetic oil and rubber were made from coal for IG-Farben. The circumstances for the Jewish prisoners in the camp were initially inhumanely heavy. The average life expectancy in the factory was three to four months; in the mines, this was only four to six weeks. Later, this became somewhat better.[10]

According to a post-war statement by Lustig, he gave a letter for Ru Cohen to a Dutch worker in the factory who was going on leave in January or February 1943. In this letter, he described 'what was going on in (his) surroundings, and that he (Cohen) had to ensure that nobody ever came here'. According to Lustig's statement, the letter reached Cohen, because he received an answer from the same worker 'that roughly read: "thanks, I have received your letter. Regards Ru."'[11]

A confirmation that the letter had indeed reached the Netherlands can be found in Hans Mogendorff's book, which is based on diary notes. He describes how he heard of the 'dramatic message' from the Amsterdam *Hechalutz* leader Kurt Hanneman 'about what went on in Auschwitz in April 1943'.[12] Hannemann considered Lustig's letter to be of such great importance that he asked Mogendorff to go to Geneva with his wife to give this letter to Nathan Schwalb of the international *Hechalutz*.

The couple intended to act as South American coffee traders. The requisite papers had to be supplied by an Amsterdam lawyer, who was also the consul of a country in that region. 'He issued passports to make the journey to Switzer-

---

[9] GFHA, Cat. no. 217, Verklaring Lore Sieskind, 5. See also Anne Frank, *Het Achterhuis* (Amsterdam: Uitgeverij Contact, 16de print, 1957), 35 (noot, 9 oktober 1942).

[10] https//www.wollheim-memorial.de., *The Number of Victims at the Buna/Monowitz Concentration Camp*, 3, accessed 14 februari 2014. See also De Jong, *Koninkrijk, dl. 8, tweede helft*, 812–817.

[11] GFHA, Cat. no. 131, Letter from Yigal Benjamin to Frits Lustig, 4 april 1988. This letter contains a reference to an earlier discussion, in which Lustig had reported these matters. Benjamin asked Lustig for a written confirmation which Lustig, however, never gave. Lustig, who lived in Deventer at the time, had little or no contact with other former pioneers.

[12] Mageen, *Van zonsondergang*, 29. Mageen (Mogendorff) writes that Lustig was a member of a so-called *Sonderkommando* in Auschwitz that cleaned trains. In this way, he could have passed the message on. This is, however, incorrect and was possibly told to him to protect Lustig and his messenger.

land possible for lots of money.' The Mogendorff's completed the necessary forms and gave passport photographs to the lawyer—but everything was in vain. A married couple that travelled just ahead of them with false papers never arrived in Switzerland. The route appeared to be unusable.[13]

In all likelihood, nothing was done with the reliable information provided by Lustig. Ru Cohen, who did pass the contents on to Hannemann, closed his mind to this sort of information or did not consider it to be credible. Hannemann may have informed the Jewish Council of Lustig's letter. However, the Council had already received this sort of information earlier. During his leave, a Dutch gardener, J.F. van Duyn, who worked for a German company in Monowitz and had spoken there with Jewish workers, told several employees of the Council what was happening in Auschwitz at the beginning of 1943. This shocked them greatly. However, to Van Duyn's great disappointment they also wondered whether he was not an 'agent provocateur'.[14]

The truth of the industrial destruction of Jews was so horrific that many people, both Jews and non-Jews, refused to believe the reports about it. Lustig's information may possibly also have reached Westerbork through rumors. But, in Westerbork there was no lack of rumors—either positive or negative, as Philip Mechanicus noted.[15]

Working and Escaping

Westerbork was not a concentration camp such as those in Germany and those in Vught and Amersfoort. Every day, hundreds of prisoners went through the gate to work outside the camp. The security was often quite light. The first job that Paul Siegel and several other pioneers had to do at the beginning of 1943 was prepare a piece of ground for the expansion of the camp's vegetable garden. During this work, he noticed that the barbed wire fence around the camp was not finished and also that the moat was still under construction. There were watchtowers with searchlights manned by the Dutch MP. However, these lights were not permanently on at night, out of fear for air attacks.[16]

Later, Siegel and other pioneers had to pluck heather shrubs to make brooms outside the camp. At the end of January 1943, a team of pioneers includ-

---

[13] Mageen, Van zonsondergang, 29.
[14] De Jong, Koninkrijk, dl. 7, eerste helft, 331.
[15] Mechanicus, In Dépôt, in many places. For an overview of what was and what was not known about the extermination camps. See also: Van der Boom, Wij weten.
[16] Abuys and Mulder, Een gat, 10. See also Siegel, Locomotieven, 92, 103–105.

ing Siegel left to *Het Apeldoornse Bos* psychiatric institution, which had shortly before been cleared of patients and medical staf. They had to help empty the buildings and pack the inventory. Other camp inmates were used in teams to dismantle Allied airplanes shot down over the Netherlands and to transport them to Westerbork, where usable components were removed for the Germans.

Besides working parties, countless camp inmates went in and out of Westerbork's gate. Some people went shopping in surrounding places. The sick went to the hospitals of Assen or Groningen for examination or admission. Sometimes, people were also given permission to go to Amsterdam or elsewhere for business or discussions. They were given a pass to get through the gate, which also gave them permission to re-enter the camp after the eight o'clock evening curfew for Jews.

In brief, there were enough possibilities to escape. Even so, it hardly ever happened. Both de Jong and Presser mention that there may have been as many as 210 escapees.[17] The number of failed attempts was undoubtedly many times higher. The German threats of reprisals on the people who remained behind formed the most important reason not to escape. Ten people from the same barracks or family members of the escaped people would be sent to the east. The former only occurred once, to the best of our knowledge, but family members and relations of refugees were certainly deported more often. The threats made a huge impression on people and were often mentioned by pioneers as a reason not to escape.

The Palestine pioneers therefore promised each other not to attempt any individual escapes. But not everyone kept to this promise. Emil Glücker (who is mentioned earlier as an *Organisation Todt* worker) walked away from the group that helped to clear out the psychiatric institution at *Het Apeldoornse Bos*. The other pioneers were very angry at him for this because he put them in danger.[18] In addition, there were other important factors, such as the remote location of Westerbork and the realization that, without papers and an address to hide, one had a great chance of being arrested again and then being sent to a concentration camp in the east as a 'punishment case'.

Yet, there were regular escapes from Westerbork. The book *Een gat in de prikkeldraad* (A hole in the barbed wire) describes several of them. Some escapes happened on impulse. Other refugees, vowing under no circumstances would they go to the east, thoroughly planned and prepared their escape. If an attempt

---

17 Presser, *Ondergang*. See also De Jong, *Koninkrijk, dl. 8, tweede helft*, 703.
18 GFHA, Cat. no. 75, Verklaring Emil Glücker, 3. See also Siegel, *Locomotieven*, 105. In his statement, Glücker says that Joop Westerweel was involved in his escape. During his work outside Westerbork, Siegel rejected various proposals to escape.

did not work the first time, they tried a second or third time. Some people got help from fellow prisoners or from outside the camp. For example, there were keys in circulation to open the door of a wagon while the train was running and to jump out. Others acted alone. Jews from the surroundings of Westerbork were sometimes smuggled out of the camp by friends or acquaintances. Several escapes happened with the cooperation of relations in the Jewish Council.

According to his own statement, the German communist Werner Stertzenbach, who, as a veteran camp inmate, had a job as supervisor of the sewerage building, smuggled 20 to 30 prisoners outside the camp in different ways. These were moreover often fellow communists, who were cared for by the underground CPN. Stertzenbach shrugged off the threats of reprisals. He argued that a maximum number of people had to be deported with every train. 'If ten people have to go with the transport for reprisal, this meant that ten others could stay.'[19]

The *Hachshara* Group in a Tight Corner

At the beginning of 1943, rumors started that Westerbork would be transformed into a work camp, with agriculture and small industries. The transports to the east would be all but stopped. The fact that as of February 1943 only one train left to Poland each week instead of two was regarded as a first sign of this new policy.

However, in the course of the year, it became clear that this was wishful thinking. Anti-Semitism was a core element of the Nazi ideology. The German measures to capture Jews intensified. In the big cities, special police units were formed to find Jews. In April 1943, a premium scheme was introduced for the arrest of Jews in hiding and thousands fell victim to these measures.[20] The reduced number of transports had nothing to do with a more humane attitude of the Germans. It was simply that more than half of the Dutch Jews had already been deported, so that fewer people were left.

The Palestine pioneers began to realize the truth that almost all Jews who were eligible would be deported to the east sooner or later. Moreover, the transport to Poland of sick and old people and children made the stories about the camps (where people would work under reasonable circumstances, for example) implausible. Paul Siegel and his friend Herman Italiaander made it a habit, when

---

19 Abuys and Mulder, *Een gat*, 22–36, 44–68.
20 For the actions of these groups, see Ad van Liempt and Jan Kompagnie, *Jodenjacht* (Amsterdam: Uitgeverij Balans, 2011).

they woke up each morning, of greeting each other not with 'good morning', but by saying, 'We are not going to Poland'. They didn't let the shocked and angry reactions of other pioneers bother them.

The new greeting was easily spoken, but quite a lot more difficult to realize— especially if you were a pioneer and wanted to stay true to the promise you made not to leave on your own. Moreover, Werner Ahlfeld, the 'Elden' leader of the *hachshara* group, was not in favor of trying to escape. Things only started moving very quickly when several of the so-called 'Palestine lists' were declared invalid at the beginning of July 1943. Forty members of the group were then put on the list for the transport to Auschwitz of 13 July. After Ahlfeld's intervention, this was cancelled.[21]

The event was, however, a sign on the wall. At the beginning of September, a list of Palestine certificate holders was again declared to be invalid. Paul Siegel describes this event as follows. In the evening of 13 September, 'the rumor spread that the Palestine-1 list had also been opened. We did not believe it. Our safety had to do with the fact that the camp commander needed us. We performed all sorts of functional work for the camp that also benefitted the inmates'. Another 'night of horrors' followed for Paul and the other pioneers. With the reading of the names of the people intended for transport, he heard 'almost immediately ... the name of someone from our group and I understood that the time had come'. However, Siegel and dozens of others did not have to go on the transports.

It is possible that all pioneers were initially selected for Poland, apart from a handful who had a named certificate. As a result of some quick work of the Amsterdam Jewish Council, approximately half could be exempted from transport. The main people responsible for this initiative were Carl (or Carli) Oroszlan and Alfred 'Zippi' Fränkel.

Carl Oroszlan was born in Vienna in 1905. After the annexation of Austria, he had been imprisoned in the Dachau and Buchenwald concentration camps for some time because of spreading anti-Nazi pamphlets. Oroszlan, whose wife was non-Jewish, so that he enjoyed a certain protection, had gone to the Netherlands at the beginning of 1939. His 23-year-old assistant, Zippi Fränkel, who was born in Breslau, arrived around the same time. After a stay in the *Werkdorp*, Fränkel went to work as a carpenter for the Jewish Council in 1941. More than a year later, he became a courier for the Council between Amsterdam and Westerbork.

Oroszlan had taken just about the same route. Through the *Werkdorp* administration, he was finally able to get a job in the Jewish Council's post room. 'The

---

21 Mechanicus, *In Dépôt*, 82, 86, 87.

lists of certificates regularly arrived from the Red Cross in Geneva with the names of people who had requested visas for other parts of the world. These people were held temporarily in Westerbork'.

Because Carl was able to open the sealed postbag for the camp, 'he could supplement these lists with other names of people who were in Westerbork'. In his estimation, about a third of the list with Palestine certificate holders was real. 'Only the first small list with the names of the so-called "vatikim" (veterans) was not false and came from Tel-Aviv.'[22]

Zippi Fränkel delivered the false lists to Westerbork. He also passed along all sorts of information about the situation in the camp to employees of the Jewish Council, so that they could take account of it. When it was known that members of the *hachshara group* had to go on transport, Fränkel passed this on to Amsterdam, from where a telegram was immediately dispatched to Westerbork with the message that the Red Cross in Geneva had issued members of the *Jeugdalijah* a certificate for exchange to Palestine.

The membership in the *Jeugdalijah* was for young people up to and including 18 years of age. Because family members of some of the other people selected for Poland reported voluntarily, the age of the Westerbork group's members could be raised to 21 years. Thus it was possible that 37 of the 77 selected pioneers remained in the camp. The 54-year-old journalist Philip Mechanicus, who was also exempted from transport at the last moment, met some of them later in the Jewish Council's office in Westerbork. '"You are certainly not yet twenty-one years old?", one of the pioneers asked him, laughing. "No not yet." "I can see that you are not." The boy then cast a look at my greying hair.'[23] Both of them had been exempted just in the nick of time.

There were forty pioneers, 12 women and 28 men—some of them just over twenty years old—who still had to go to Auschwitz on 14 September 1943. They and the people who remained behind turned the departure into a demonstration of Jewish self-awareness and pride. The entire *hachshara* group accompanied the 40 pioneers to the train. The people who remained behind were supposed to go back to the barracks. However, they crept outside, to be as close as possible to the platform. 'The train doors closed, the parting hit us hard. We sang *Hatikva* (the Hope) together and *Techezakna* (Stay Brave). My voice stuck in my throat and many of us cried freely.'[24] The departure of the pioneers left a lasting impression on everyone present.

---

[22] GFHA, Cat. no. 165, twee ongedateerde Verklaringen Carl Oroszlan, 1, 2 en 1–8.
[23] Mechanicus, *In Dépôt*, 159, 160.
[24] Siegel, *Locomotieven*, 118.

Gemmeker gave the selected people a letter 'in which he pleaded for the group to continue the same sort of work that they had been doing in Westerbork i.e. work in the fields, and in the same groups'. In a cynical way, the Germans also did this to a degree. The men ended up in the Buna-Monowitz complex. Here, they met Frits Lustig, who asked them indignantly: 'What are you doing here? I wrote to Ru that nobody should come here.'[25] The 12 women were locked up in the medical experiments barracks of Auschwitz. Ten of them would return; of the 28 men, only two did.

Organized Escapes

The stark realization that nobody was in fact safe from deportation from Westerbork led the day after the departure of the 40 people to the escape of pioneer Lily Kettner, whose time spent hiding in Sevenum is described earlier. Lili was hardly an obvious candidate for deportation. Her parents lived in Palestine and were therefore British citizens. For this reason, Lily had been issued with a *Sperre;* a Palestine certificate was underway. She was arrested in June 1943 in a large raid in the center of Amsterdam, sent to Westerbork and placed with the *hachshara* group. The transport of 14 September made her decide to flee the next day. She previously had talked about a possible escape with Erika Blüth of the *Jeugdalijah* and Mirjam Waterman. Because of the chance of reprisals, they had then decided not to escape.

Lily worked in a laundry in Meppel, which was a great advantage. At the end of the working day, Mirjam waited for her at the station and gave her 'a paper on which officially stood that I was allowed to go to Amsterdam for a few days to arrange my (certificate) affairs'. She also got a train ticket for Amsterdam. Lily then boarded the train to Assen, 'went to the toilet, went out of the toilet to the opposite platform and boarded the train to Amsterdam. I got out of the train in Zwolle with a star on my coat. All sorts of passers-by wished me good luck, and I stepped in the train to Amsterdam. Mirjam and I found an empty train compartment. Mirjam had another coat for me and some photographs of me that she had picked up somewhere in Amsterdam. I changed my coat and we made an identification card with these photographs and a real thumbprint and I wrote my new name on it: Aafje Hoving'.[26] After some time spent searching around for a place to hide, Chiel Salomé finally took Kettner to Sevenum.

---

25 GFHA, Cat. no. 131, Brief Y. Benjamin aan Lustig, 4 april 1988.
26 GFHA, Cat. no. 111, Interview Mirjam Pinkhof-Lily Kettner-Ofek, 1987, 2.

Lily had gotten away safely and, possibly even more importantly, there were no reprisals against other pioneers—who were, however, prohibited for some time from working in the laundry in Meppel. This stimulated the *Hechalutz* leaders to set up other escape attempts. There is no complete overview of these attempts. What is known is that Mirjam Waterman and Lore Durlacher were initially involved in these escapes and later also people such as Jan Smit, Frans Gerritsen and Tinus Schabbing. Two non-Jewish girls, who were friends with pioneers, became friendly with members of the Dutch MP to get their cooperation. Several people from Drenthe who lived in the surroundings of the camp also lent a hand.

Pioneers David Rosenbaum and Martin Uffenheimer, who worked in the farming group outside the camp, escaped, probably shortly after Lily Kettner. They left their work and walked to Beilen. Near the station they met an officer of the Dutch MP who recognized them. 'We tried to appeal to the officer's conscience but nothing helped. We were sent back to the prison barracks in Westerbork'. They were spared immediate transport to Poland through the action of Esther Singer, a friend of Uffenheimer. She had a job as a maid in Gemmeker's house, and asked him if he might allow the two young men to rejoin the *hachshara* group. Almost nobody succeeded in getting out of prison, but this time Gemmeker agreed.[27]

The fact that Rosenbaum and Uffenheimer could stay in the camp was a good thing. However, their unfortunate experience clearly emphasized the dangers posed by indifferent or hostile surroundings. Several other events also made it abundantly clear that it was time to flee from Westerbork and not to trust in lists or in German promises. With regard to these promises: the Jewish Council was dissolved at the end of September 1943, after which almost all employees were sent to Westerbork. The same happened to the more than 600 Jews of the so-called Barneveld group, who lived in a small castle in the village of Barneveld. They had been identified by the Dutch authorities as worthy of protection, in recognition of their contributions to society. The Germans had promised not to deport them—but this was now going to happen anyway.[28]

---

27 GFHA, Cat. no. 191, Informatie David Rosenbaum in Brief Moshe Keny aan Jan Smit, 21 juli 1994.
28 In 1944, the Barneveld group was transferred to the Theresienstadt concentration camp, which had a relatively good reputation.

A Failed Mass Escape

It was clear to the *Hechalutz* and the Westerweel group's helpers that a large-scale escape attempt was required to get the vulnerable Palestine pioneers to safety. A passport photograph was made of them in Westerbork, probably in September. This happened under the guise of setting up a sport club, for which the members needed a membership card with a photograph. More than 120 false identity cards were made in Amsterdam with these photographs, which were stored by Werner Ahlfeld, the leader of Elden group. Although he was above 21 years of age, he had been able to stay in Westerbork because his wife was on a 'good' Palestine list.

At the beginning of October 1943, Ahlfeld took several pioneers with him to Amsterdam to clear out the deserted offices of the Jewish Council. During his stay there, Menachem Pinkhof contacted him about a plan for the escape of the entire *hachshara* group from Westerbork. The intention was that this plan (in which Joop Westerweel was probably also involved) would be a larger scale repetition of Loosdrecht, where pioneers had all left on one night to hide at addresses spread throughout the Netherlands.

The massive escape plan meant a breach with the line followed up to then that escapes had to happen in a way that no reprisals would follow. Menachem Pinkhof had had a prior talk with the prominent Jewish Council employee Kurth Blüth. Kurth had told him 'that, despite everything, it was responsible to get as many *chaverim* as possible out of the camp, no matter how'. His words had given Pinkhof the idea that Blüth knew more than the others about what went on in the Polish camps.[29] This was also almost certainly so.

However, the plan was of necessity more complicated than the flight from Loosdrecht, because they now also had to get past the camp guard. Westerbork's isolated location was also a hindrance. To leave the camp, they obtained the co-operation of a Dutch MP, who was ready to guide the pioneers through the barbed wire 'for a large amount of money'. Ten pioneers (two women and eight men), including Herman Italiaander, Martin Uffenheimer and Paul Siegel, were given the task of cutting through the wire and leading the others outside. There, one or two stolen German army trucks would be ready to transport the escaped pioneers to a hiding place, from where they would go to different hiding addresses. According to Ahlfeld, one day before the mass escape, it appeared

---

[29] GFHA, Cat. no. 173, Verklaring Pinkhof, biographical note, 11, 12. See also Siegel, *Locomotieven*, 212, 213.

that 'it was impossible to execute the plan'. The telephone code message for calling off the plan was: 'No suitcases will be sent'.[30]

Probably shortly afterwards, Italiaander and the others of the wire-cutting team (thirteen men in total) attempted to escape, again with the cooperation of a Military Policeman. He was an acquaintance of Tineke de Lange, Carl Oroszlan's girlfriend. Frans Gerritsen and Lore Durlacher would cut through the barbed wire from the outside, just after a guard patrol had passed. They were connected by means of a 500-meter-long wire to Tineke, who had agreed with her MP acquaintance that he would keep an eye on the situation in the camp. If there was any danger, he would give a red light signal. If everything went well, the escapees would creep along the wire over the moor to Frans, Lore and Tineke, after which they could flee further.

The escape did not go through, however. When Frans and Lore lay in wait at the barbed wire, Frans recalled later, 'I got a tug on the foot, so we went back and large searchlights immediately shone over the moor. So, something had gone wrong, we did not know what. We had black clothing on and stayed still. When the light had gone, we crawled back along the rope. Tine said yes, that Dutch MP had signaled with a red light. Something had happened inside but we did not know what.'[31] Italiaander and the other twelve pioneers spent a cold November night in Westerbork on the moor, but were able to return to their barracks.

A Train Driver as Coordinator

After the failure of the larger-scale escapes, the *hachshara* group was forced to return to smuggling one or more individual pioneers outside the camp. In Westerbork, the co-ordination of these escapes was mainly in the hands of Kurt Walter, who had a special position in the camp. He was namely a train driver on a narrow-gauge train with tippers (dump cars), which transported goods 5 kilometers to and from the Oranjekanaal.

Kurt Walter was born in 1922 in the small town of Bamberg, in the north of Bavaria, where his father had a company in agricultural machines. After he was imprisoned in the Dachau concentration camp in November 1938, Kurt went to the Netherlands. He worked in the *Werkdorp Wieringermeer* until March 1941. After its closure, he became an assistant teacher at the M.S. Nijkerk Ambachts-

---

30 GFHA, Cat. no. 1, Verklaring Ahlfeld, 3.
31 GFHA, Cat. no. 63, Interview Yigal Benjamin-Frans Gerritsen, 29 december 1989.

Fig. 22: Kurt Walter, shown here on his locomotive, was the main organizer of the escapes from Westerbork

school (technical school), which was managed by the Jewish Council. He was arrested during a large razzia at the end of May 1943 and taken to Westerbork, where he became a member of the *hachshara* group.

Kurt initially worked as a supervisor on an expansion of the camp. When a narrow-gauge railway was laid, his first job was to repair the small locomotive. He then went to work as a train driver. In mid-July 1943, he was sent to Amsterdam to dismantle machines in the Nijkerk School, whose pupils had been deported. Lore Durlacher approached him to act as a contact of the *Hechalutz* in the camp. Initially, he did this together with Zippi Fränkel, the courier of the Jewish Council, but Zippi went into hiding when the Council was closed.

On his journeys to the Oranjekanaal, the rails passed through a wood. There, Walter would stop his train to meet with Lore or another helper. He was then given false identification cards and other documents, plus the names of the candidates for escape and also keys to open train doors and tools to aid in the escape. In the camp, he gave these items to *hachshara* group members who were engaged in the resistance, such as Paul Siegel and Herman Italiaander. Werner Ahlfeld and others of the Elden group, who were considered to be unwilling to participate in the escapes, were kept unaware of the plans.

In a 1955 interview, Kurt Walter described the organization of the flight of the pioneers from Westerbork as follows:

> We tried to organize the escape such that no reprisals could be taken against other camp inmates. Therefore, we had to wait until the relevant *chaverim* were on the transport list. These *chaverim* would have to leave the line of prisoners on their way to the train and go to the latrine unnoticed. After the train left the camp, they would go to a previously-agreed place.
>
> There they would spend the night, and leave the camp the next morning with forged identity cards. Outside, *chaverim* of the illegal organization would wait for them at an agreed place. The escapees would be taken by bicycle to various places in the Netherlands to go into hiding. Most of the escapees would be provided with forged identity cards while they were still in the camp.[32]

Of course, the escapes did not all go according to this blueprint. For example, using the latrines as a hiding place appeared to be no good because they were checked. However, the above text gives a good idea of the plan. The escape plans that were developed in October 1943 were initially postponed for some time because several cases of polio were ascertained in Westerbork at the beginning of November. After the transport of 16 November, no more trains left for Poland because of the danger of infection.

At the beginning of January 1944, however, notice came that the transports would be resumed. The first train did not go to Poland, but to the Bergen-Belsen camp in Celle on the Lüneburgerheide, about 60 kilometers from Hannover. This camp was regarded as being reasonably good. Here, there were Jews with neutral nationalities, Jews with nationalities of Latin-American countries and Jews with a Palestine certificate. The powers in Berlin hoped to be able to exchange this last group for Germans from countries of the British Commonwealth and Palestine.

Forty members of the *hachshara* group went to Bergen-Belsen, by passenger train. The next train went to the 'model camp' of Theresienstadt in Czechoslovakia, which also had a good name. However, the following train went to Auschwitz. The transport on 1 February again had Bergen-Belsen as its destination. Paul Siegel and Manfred Uffenheimer reported voluntarily for this. However, they were determined to be the first to escape according to Kurt Walter's blueprint.

Apart from the fact that Siegel injured his knee when he jumped out of the train at the last moment before it left, everything went according to this plan. The

---

32 GFHA, Cat. no. 251, Verklaring Kurt Walter, oktober 1955, 1–4.

two hid in the hospital barracks and later in the horse stables close to the gate. In the early morning, they left the camp with a false pass and walked to a wood with dense bushes about a kilometer away. Later, because of the rain, they went to a haystack at a farm, where they hid during the day. In the evening, they went to a meeting point at the Oranjekanaal, where Lore Durlacher and Frans Gerritsen stood with extra bicycles. They had kept hidden in a turf-hut nearby, which was discovered coincidentally by Lore. The hut had been made by three Drenthe people in hiding for the *Arbeitseinsatz*. Jan Smit had agreed with them that the Westerweel group could also use the hut. It was a good solution for the damp and cold wintry weather on the Drenthe moor. The four of them cycled to Assen, where they spent the night with a young family.[33]

The next day, they travelled separately by train to Amsterdam. However, before they arrived, they had a few anxious hours caused by a delay of the train at the station of Assen. The station was full of the German soldiers, police and Dutch MPs. This was not much fun for Siegel and Uffenheimer, who both had a Jewish appearance and were well known in Westerbork. But it was also a good lesson for the future: panicking or showing fear was simply not an option.

In Amsterdam, they stayed the night with Lore, and were introduced to Jan Smit. He said that a contact in the Amsterdam police had told him that raids were planned on Jews and students in hiding. Therefore, it was better to go into hiding in The Hague with Pim van Maanen and Gerda Loeff, who had got married shortly before. They were former pupils of Joop Westerweel at Kees Boeke's *Werkplaats*. After the *Werkplaats*, Gerda Loeff, who was born in 1922, had followed a course in graphic arts at the Haagse Academie. She had been imprisoned for two weeks in May 1942 because she and a friend of hers had worn a Jewish star as a protest. Pim van Maanen, who was two years older and known later as the writer Willem van Maanen, worked as an art critic with the *Algemeen Handelsblad* newspaper.

Pim and Gerda had remained in contact with Westerweel after their school time and, in the autumn of 1943, they asked him if he could use some help with his resistance activities. He had then explained the dangers and given them fourteen days to think about it. When they decided to move forward, Gerda was given the job of helping Frans Gerritsen with forgeries. In February 1944, they started hiding Jews, although their home was not really suited for this. The owner of the house, who lived on the ground floor, was namely a member of the NSB. His son, also a party member, had a room next to Gerda and

---

33 *Emmer Courant*, Schiedammer zoekt oude oorlogsvrienden; Onderduikers leefden anderhalf jaar in hol, unknown date 1984. See also Siegel, *Locomotieven*, 127–140.

Pim's apartment. By keeping as quiet as possible, Siegel and Uffenheimer were able to stay in The Hague for about ten days.[34]

One evening, the doorbell rang, and the two pioneers were called to come out. Paul Siegel later wrote in his memoirs that they saw a man who can best be described 'as a travelling salesman' with a short haircut, spectacles and a little moustache. Gerda introduced him as Joop Westerweel, about whom 'was spoken with much respect' because of his resistance activities. He had been described to Siegel as a 'youthful looking person with a thick head of hair', but Joop told them that he had to change his appearance because the Germans were trying to catch him after the arrest of his wife in December 1943.

Shortly afterwards, Gerda Loeff took the two young men to Eindhoven by train. She was a pretty blonde woman with an open manner who naturally drew attention to herself, so that her Jewish travelling companions could travel in peace. In Eindhoven, Paul Siegel and Manfred Uffenheimer stayed the night in Jasper Daams's house. The next day, Jan Smit and Joop Westerweel took them across the Belgian border at Budel as a first stage on the way to France. After several days in Brussels, where pioneer Riwka Roos joined them, they left for Paris, accompanied by 'Cor' (Max Windmüller).[35]

Following Escapes

Herman Italiaander and Frits Siesel were the next two pioneers who fled from Westerbork. Their escape went more or less according to the pattern of their predecessors. They volunteered for a transport to Theresienstadt on 25 February 1944 and got out of the train at the last moment by quickly putting armbands on that belonged to members of the porters' service who helped the deportees. Herman and Frits left Westerbork the same day with the aid of a false pass and spent the night in the woods outside the camp. The next day, Lore Durlacher and Frans Gerritsen picked them up with two extra bicycles. They went directly by train, via Assen, to Pim and Gerda van Maanen's home in The Hague. A week later, Joop Westerweel appeared and took them to Belgium.[36]

The next group that escaped was comprised of six pioneers and had a somewhat different background. Most of them had only been in Westerbork for a short time after they were arrested at the beginning of 1944 in an attempt to flee to

---

[34] Interview Sytske De Jong-G. Loeff, 25 oktober 1998, 7–13.
[35] Siegel, *Locomotieven*, 142–149.
[36] Kochba and Klimov, Manuscript, hfdst. 6, Bijdrage Herman Italiaander, 249–250.

France. Their status as 'punishment cases' was then changed and they became absorbed in the *hachshara* group. However, in mid-March, together with 23 other pioneers, they were told that they had to go on transport to Auschwitz. Initially, ten and later six of them decided to escape. Everyone had to wait in line for the transport, and the authorities would check off their name from the list. When the following words were written after their name: 'Added to the transport', they disappeared quietly from the row that walked to the train and hid in the school barracks. They had agreed beforehand that they would report immediately if their flight was discovered, to avoid reprisals.

> The room is full of cigarette smoke. In our nervousness, we light one cigarette after the other. We keep looking at our watches. At half past nine, the wagons are closed and the train leaves. The time passes slowly. The tension becomes unbearable ... A whistle sounds. We listen. Then we hear the hissing of the locomotive. Slowly, 25 cattle wagons with 800 Jews in them roll out of the camp.[37]

The escapees did not have to report. Using false passes, they left the camp after the lunch break in two groups of three. They spent the night in the woods, where they built a makeshift hut with branches. After the last instructions from Kurt Walter to stay the night in the hut and to be at a certain place at exactly 6.45, they spent a heavy night with hard wind and heavy snowfall. At 6.43, the refugees were at the agreed place. 'Sitting behind a small hill for two minutes, we then hear footsteps. We immediately distinguish two figures trying to find their way—a woman and a man. With a muffled cry of joy, we fall into each other's arms. It's Lore and Frans.' Later, Jan Smit joined them as well. There was a total of eight bicycles, so that everyone could get away. Most of this group would also travel through to France.

There is no complete picture of the escapes. It is also not clear how many escapes from Westerbork were made. According to his own claims, Frans Gerritsen was involved in 16 escapes. Martin Uffenheimer, an accurate reporter, mentions a number of approximately 21 escaped pioneers. Jan Smit reported 30 escapes. However, this appears to be too high. A number of around 25 escapes from the camp is the most realistic.

Almost all escapes ran according to Kurt Walter's blueprint, with some variations. But not everything went smoothly. In one instance, there was a transport of two young escaped pioneers (a boy and a girl), who Lore Durlacher and Frans

---

37 GFHA, Cat. no. 93, Verklaring Werner Hirschfeld, augustus 1955, 1–5. See also Kochba and Klimov, Manuscript, anonymous contribution, 252–254. Hirschfeld was arrested at the beginning of 1944 in the earlier-described failed transport of twelve men with false leave passes, in which the German deserter was arrested and later executed.

Gerritsen transported on the backs of their bicycles. They were arrested close to the camp by a Dutch policeman. He 'stopped them and asked where they were going to. Lore answered that they were looking for food. The boy, (on the back of Lore's bike, hs) did not have an identity card. Lore saw that the police did not believe them and then tried to appeal to his humanity. She said that she wanted to save her fiancée from the camp'. Lore, who wrote the report, gave the pioneer her bicycle and remained behind with the policeman. She was successful in convincing him not to take any action. She walked to the agreed meeting point, met Frans and the pioneers and travelled with them to Amsterdam.[38]

In the case of unexpected transports, which sometimes occurred, one had to improvise. This was the case, for example, with one of the last escapes at the end of March or the beginning of April 1944. It had only become clear during the evening that a transport would go the next day. Several helpers appeared unreachable, but Tineke de Lange, who had taken over the co-ordination, found Tinus Schabbing willing to go to Westerbork with her. After an especially cold night that he and Tineke spent in a chicken coop, they succeeded in helping former *Werkdorp* pioneer Julius Reutlinger and his wife Hella to escape and get to Amsterdam.[39]

Attempts at Sabotaging the Transports

The escapes from Westerbork ended in April 1944. During a roll call in Bergen-Belsen after the arrival of a transport at the beginning of February 1944, it appeared that Manfred Uffenheimer and Paul Siegel were missing. However, there was no direct link with Westerbork because they could also have fled somewhere on the way. In addition, however, the Germans got information about the escapes from the infiltrator 'Karel' in March or April (see also chapter 10), who had stayed in Pim and Gerda van Maanen's house for some time. He had spoken extensively there with Paul 'Amo' Rosenbaum, the pioneer who had escaped from Westerbork, who had told him all sorts of things about his flight.[40] After this, the camp guard was intensified. The Germans searched the 'escape wood' close to the camp and also found the turf hut.

With the help of Kurt Walter, at least one pioneer escaped from Westerbork later. This was Gideon Drach, who had been imprisoned in Vught for about a year

---

38 GFHA, Cat. no. 45, Interview Lore Durlacher, augustus 1955.
39 Interview Sytske De Jong-Tinus Schabbing, 13 oktober 1998, 13, 14.
40 Abuys and Mulder, *Een gat*, 42. See also GFHA, Cat. no. 242, Verklaring Manfred Uffenheimer, april 1956, 3.

Fig. 23: Lore Durlacher cared for pioneers in hiding and was involved in many escapes from Westerbork. The photograph is from 1942.

after his arrest in August 1943. He was sent from Vught to Westerbork as a 'punishment case' and selected for transport at the beginning of September 1944. Walter hid him in one of the tippers of his narrow-gauge train. Drach fled to a nearby farmer who could be trusted, and then to Assen, where he hid. He was betrayed, arrested again in the beginning of 1945 and locked up in Westerbork, where he stayed until the liberation.[41]

---

41 GFHA, Cat. no. 70, ongedateerde Notities Kurt Drach.

At the beginning of September 1944, the rumor spread in Westerbork that the camp was dissolved. All inmates would be transported to Poland. An alarmed Frans Gerritsen decided to intervene by blowing up the rails near the camp. He contacted the *Raad van Verzet* (RVV, Resistance Council) and asked them to make a sabotage team available. He travelled to Assen with three men, but it appeared that they had no explosives with them and there was not enough time to obtain them in Assen. Frans and Lore ceased all further cooperation with this group, 'because two of the three RVV people were sensationalists', who they preferred to avoid.[42]

Together with 'Klaas', a Drenthe resistance man, Lore and Frans discussed what could still be done to sabotage the railway line. It was decided to let a normal train derail at the points where the railway line went to Westerbork; this would also block this line. One of their acquaintances in Assen had a heavy iron vice that they could use could for this purpose. 'We picked the thing up in the very early morning and went to the railway line. Naturally, the line was guarded. When the vice rattled a bit at a railway crossing, the guard looked around, stood still for a moment and continued on his way along the endless tracks.' During his bike ride 'in the mucky drizzle' to Westerbork, Frans wondered if what he wanted to do was ethical, 'simply (to derail) a passenger train that happened to pass these points, just as a protest? I don't know any more ...'[43]

However, he carried on and was quickly confronted with the fact that the open railway embankment, which offered no cover at all, was heavily guarded and moreover within sight of a signal box. He left the vice behind and climbed onto the embankment to reconnoiter the situation. At that moment, two Dutch MPs appeared from the direction of Westerbork on bicycles. They looked in his direction but did not see him—or did not want to see him.

After this shock, Frans would preferably have left immediately. But he decided to try again. You could do nothing and, 'if everyone else says that this war is lasting too long and the bread is so bad; or that we should do something good or bad as a protest, it could not bother me anymore, but something had to happen'.

He picked the vice up and 'after looking in all directions', he quickly climbed on to the railway embankment. 'The jaw (of the vice) bit on the rails. I nervously tugged on the vice until it got tighter and tighter. When it was finally tight, I kicked the handle a few times and it was fixed unmovable on a fish plate.'

---

42 GFHA, Cat. no. 63, Frans Gerritsen, Brief 16 september 1945. In this letter, which was intended for the Westerweel group's members, Gerritsen gave a report of his activities from the autumn of 1944 to May 1945.
43 Ibid.

Frans cycled quickly away, but saw that a patrol of seven Dutch Nazi policemen were coming.

In a café in a village nearby, Frans waited for Lore, who had gone to check the situation. She said that the train that should have derailed had arrived without delay. 'Everything had therefore been for nothing and the injustice went on! The vice was probably run over and destroyed, or those seven policemen found it.'[44]

As far as can be ascertained, this was only one of the very few attempts to sabotage the transports from Westerbork—and it ended in failure. At the same time, however, one could call Frans Gerritsen and Lore Durlachter's attempt a triumph of the resistance against anti-Semitism and the general indifference to it in the Dutch population.

Westerbork was, moreover, not cleared out. The last transport left on 13 September 1944 to Bergen-Belsen. As a result of Operation *Market Garden* (the Allied air landings at Arnhem that were intended to liberate large parts of the Netherlands), a general railway strike ordered by the Dutch government in London started mid-September, after which no more trains left the camp.

When the rumor spread in the spring of 1945 that Westerbork would be cleared out and the inmates would be transported or liquidated, Frans Gerritsen and Kurt Walter thought up a plan to save several of them in any case. Kurt knew of a 200-meter-long sewer pipe that went under the moat and the barbed wire to a purification system outside the camp. The pipe's cross-section was approximately 50 cm. It was possible to crawl through it—and there was sufficient oxygen, as Kurt Walter had demonstrated with a burning candle. Frans therefore constructed a wagon with a length of about 1.5 meters, on which someone could lie and be drawn through the pipe.[45]

Luckily, this plan never needed to be tested. The Germans fled before the Canadians liberated the camp on 12 April 1945. There were then still almost 900 Jews in Westerbork, including Kurt Walter and Gideon Drach.

The Palestine Pioneers in Bergen-Belsen and Other Camps

At the end of November 1943, the *hachshara group* in Westerbork consisted of 93 people. Nine of them came from the ultra-orthodox *Agudah* kibbutz in Enschede, 32 from Elden, three belonged to the *Jeugdalijah group* from Loosdrecht, 28 came

---

44 Ibid.
45 Ibid.

from Deventer and the Werkdorp, 17 belonged to the *Mizrachi* (the religious *hachshara*) and four came from Gouda.⁴⁶

This number will have changed afterwards as a result of new arrests and escapes, but it can be assumed that more than 90 pioneers were in Westerbork in the spring of 1944. The majority, approximately 75, were deported to Bergen-Belsen in the first months of that year. About 15 went to Auschwitz and several others to Theresienstadt, from where they were also sent to Auschwitz, in violation of all promises.

In Bergen-Belsen, the *hachshara* group's members were put in the so-called *Sternlager* (Star camp), where initially mainly Dutch Jews stayed who the Germans hoped to be able to exchange for other Germans in Allied hands or for goods. The group strove to continue its activities from the *hachshara* training and Westerbork. There were lectures and discussions—and the study of Hebrew was resumed. The group also organized education for children in the camp.

Although there was no obligation to work at first, the Palestine pioneers volunteered for work. Most were put on kitchen duty and were given jobs such as peeling potatoes. Others had less luck and had to remove the leather from old shoes in a large tent or dig out tree roots outside the camp, which was unhealthily dusty or very heavy work, respectively. Some people were lucky enough to be able to work in their old professions. Werner Ahlfeld, for example, was made head of the technical service in the *Sternlager*, together with another prisoner.⁴⁷

Between 11 January and 13 September 1944, more than 3,700 Jews were deported from Westerbork to Bergen-Belsen. Most of them (about 3,000) were deported in five transports between 11 January and 5 April. Of these people, approximately 450 had a Palestine certificate, which meant that the British government could allow them into Palestine. In June 1944, after the usual red tape, 222 of them (110 Dutch and 101 German Jews plus 11 Jews with passes from a Latin-American country) were exchanged in Istanbul for 150 Germans from Palestine and South Africa. Six people in this transport were young women from the original Elden group.⁴⁸

The circumstances in Bergen-Belsen were initially relatively favorable. The camp appeared much like Westerbork, both through his location on the Lünebur-

---

46 Samson, *Zwischen*, 204, 205. The list contains 93 typed names, with 14 names of pioneers who arrived later in Westerbork added in handwriting.
47 Kochba and Klimov, Manuscript, hfdst. 6, Bijdrage Lotte Levi-Silmann, 254–255. See also GFHA, Cat. no. 1, Verklaring Ahlfeld, 4.
48 De Jong, *Koninkrijk, dl. 8, tweede helft*, 736–745. See also GFHA, Cat. no. 1, Verklaring Ahlfeld, 4. More than another 1,800 mainly Hungarian Jews reached Switzerland in three later transports.

gerhei and as a result of its organization, which had a certain degree of self-government. However, the situation gradually worsened. An obligation to work was introduced, with working days of at least 11 hours, while the amount of food was gradually reduced. There were often roll calls that lasted for hours and large groups of prisoners were repeatedly given no food as a punishment.

From the end of 1944, the camp was flooded with prisoners from concentration camps in the east, which had been evacuated because of the Red Army's advance. This created a chaotic situation, which had catastrophic consequences due to the failing food supply and outbreaks of diseases such as typhus. When British troops liberated Bergen-Belsen on 15 April 1945, they found thousands of unburied corpses. Of the 60,000 prisoners who were still alive, 13,000 died in the days and weeks after the liberation. Of the 3,750 prisoners who had arrived from Westerbork about eight months to a year before, about 1,700 died.[49] The *hachshara* group managed to get through the stay in Bergen-Belsen reasonably well. A very important aspect here was that a number of members were able to keep their work in the kitchen. They shared the extra food that this work gave them with other pioneers. Moreover, they were young and hardened by the heavy work in the *hachshara*.

Yet there were victims. Manfred Samson's memoirs show that eight pioneers died in Bergen-Belsen. These included the pioneers Hans Horowitz and Arnold Koller, who originally came from Loosdrecht. They were put to work in the camp crematorium. In November 1944, the Germans took them to the Neuengamme concentration camp in Hamburg, where they were murdered. Hans Stein, a 'punishment case' from Westerbork, was betrayed in Bergen-Belsen and sent to Auschwitz, where he died.[50]

At the beginning of April 1945, the Germans decided to evacuate the *Sternlager* and to send the roughly 6,700 prisoners to Theresienstadt by train. The first train left on 6 April, but did not reach its destination. The passengers, including several Dutchmen, were liberated by the Americans near Magdeburg three days later. A second train with mainly Hungarian Jews left on 7 April and arrived in Theresienstadt two weeks later.

The third train, which left Bergen-Belsen on 11 April with more than 2,500 prisoners, carried many Jews from Westerbork, including a large part of the *hachshara group*. Theresienstadt appeared unreachable and the train wandered through the east of Germany for fourteen days—heedless of the wellbeing

---

49 De Jong, *Koninkrijk, dl. 8, tweede helft,* 845.
50 https//www.its-arolsen.org., accessed 14 juni 2014, The Liberation of the 'Lost Train' near Trobitz, 1, 2.

of the passengers. The passengers were liberated by the Russians on 23 April 1945 at the small town of Tröbitz near Leipzig. In the meantime, 200 prisoners had died. After the liberation, another 320 passengers died from illnesses and exhaustion. According to Samson's notes, there were four deaths among the *hachshara* group's members in the train and in Tröbitz. More than 60 of the approximately 75 Palestine pioneers who stayed in the camps survived this last phase of the war.[51]

---

51 Samson, *Zwischen*, 204, 205.

10 The German Counter-Offensive

This chapter investigates how the Germans reacted to the Westerweel group's activities. This is easier said than done. Almost all of the archive material of the German police services was destroyed at the end of the war. It is therefore not known how they assessed the first resistance initiatives of the Westerweel group in August 1942. What we do know is that the reprisals feared by Menachem Pinkhof and Shushu Simon after the Loosdrecht group went into hiding never materialized. The Germans possibly regarded the disappearance of the Loosdrecht group as a sort of accident.

Infiltrations and Arrests

There is more known about the measures taken by the German police in the autumn of 1942. These were aimed at collecting information about the activities of the *Beth Chaloets* leaders in the Tolstraat. It is not clear to what extent their efforts were linked to the disappearance of the Loosdrecht group. There were one or two raids on the pioneers' house, whereby the inhabitants were interrogated. Two female pioneers hiding there were arrested and taken to Westerbork. The Germans will have formed a picture of the activities of at least the Jewish part of the Westerweel group from the interrogations. At the end of 1942 or the beginning of 1943, there followed the efforts of the infiltrator 'Mrs. De Ruiter', who lived in Amsterdam. Her name was sometimes also written as 'De Ruijter', but she was really Maria van Ginkel.

Together with her boyfriend Dirk de Ruiter, she had a guesthouse in the Dintelstraat in the Rivierenbuurt in Amsterdam, that had a high Jewish population. 'Mrs. De Ruiter' worked closely with the Utrecht Kindercomité (children's committee), which hid Jewish children, but she also maintained contacts with the German police. She had come into contact with the *chalutzim* through some pioneers from Deventer who stayed in Amsterdam. It is not known how this contact came about.

One of them told *Hechalutz* leader Kurt Hannemann that 'she wanted to help us with our resistance work'. Norbert Klein then assessed 'Mrs. De Ruiter' as suitable, and became her contact. She supplied money and goods to the organization, visited the *Beth Chaloets* and even had two pioneers under her care. After her exposure in March 1943 through a warning in the illegal newspaper *Vrij Nederland*, Klein broke all links with her. The pioneers left the Tolstraat with their possessions; the two people in hiding for whom 'Mrs. De Ruiter' was caring were quickly picked up by the Westerweel group.[1]

The reaction was predictable. There was another raid, now by six agents of the *Sicherheitspolizei (SiPo)*. There were interrogations but no arrests, for the moment. The interrogations revealed that the Germans were really searching for Norbert Klein, whose identity, however, they did not know. He was advised to stay away from Amsterdam for a time and received a false identification card.

Klein, a modest, withdrawn person, who had carried out countless courier's assignments with great courage, became very upset by all this attention. He started to smoke, dressed conspicuously and told others that at least six German agents were hunting him. He was in fact no longer suited for resistance work and the best thing for him would have been to go into hiding. However, his efforts could not be missed. At the beginning of August 1943, he was arrested with incriminating material. After a first interrogation, he jumped out of a window on the second floor of the police office in the Paulus Potterstraat the next day.[2]

Norbert Klein was badly wounded and the Germans took him to the guarded ward of the Wilhelmina-Gasthuis (hospital). Frans Gerritsen heard of Klein's arrest from Kurt Hannemann and Lore Sieskind, who were living with him and his wife in Haarlem. He contacted a nurse in the hospital who was friends with Joep Nubert, the leader of a resistance group. Together with Gerritsen, they worked out a plan to liberate Norbert. At the beginning of September, when Norbert was getting better, he was 'lowered from the window on a stretcher with incredible speed and dexterity and slid inside a Ford car through the open roof, practically under the eyes of the guards.'[3] Norbert Klein, who had a serious back injury, was hidden in the Gerritsen family's house, where he was cared for by a doctor.

The day after Klein's arrest, the Germans also arrested Gideon Drach, the other *Hechalutz* courier. He had gone to Klein's room with Palestine pioneer Helga Decker to remove incriminating material. However, Klein's landlady, who Drach did not trust, did not want to let them in. The police, who had possibly been warned by the landlady, followed Gideon and arrested him in his

---

[1] GFHA, Cat. no. 33, Verklaring Leo Cohen, januari 1957, 2, 3. Cohen was born in Enschede in December 1916. As a result of the betrayal by 'Mrs. De Ruiter', Bella Schenker and Otto Sluizer and others were arrested. No further information is known about Schenker. Sluizer was murdered in Bergen-Belsen in 1945. See also *Vrij Nederland*, 21 maart 1943, Rubriek waarschuwingen. After the warning, Van Ginkel and De Ruiter moved to St. Michielsgestel in Brabant, where a resistance group from Utrecht attacked them. Dirk de Ruiter was killed; his girlfriend severely wounded.
[2] Mageen, *Van zonsondergang*, 56, 58.
[3] GFHA, Cat. no. 63, Verklaring Frans Gerritsen, samenvatting nadere mededelingen, ca. 1957, 3.

room. He was interrogated and kept prisoner for six weeks in the Amsterdam *Huis van Bewaring*. He was then transported to Camp Vught, where he had to unload trains with aircraft components on the station platform. Later, he worked with the so-called *Philips Kommando*. (This was a group of prisoners that worked for the Philips Company in the camp).

Two attempts were made to free Gideon Drach. Joop Westerweel visited him once at work on the station. He had false papers, civilian clothing and a cap (for Drach's clean-shaven head) with him and told Gideon to change his clothes and get into a departing train. However, Gideon found the risk too great and stayed in Vught. Later, Frans Gerritsen (who had obtained access to the camp as an 'engineer') arrived with false papers. He suggested bribing the *kapo*, the head of the working party, to let Gideon escape. Gideon agreed with the plan, but the *kapo* did not accept the offer.

Because he was registered as a 'half-Jew', Drach once at the last moment was exempted from a large transport from Vught to Westerbork. Finally, he lost this status and was transferred to Westerbork anyway.[4] Although he was selected for transport to Poland, he escaped, as mentioned in the previous chapter, with the help of Kurt Walter.

The Safe House in the Dordtsestraatweg

The arrest of the two most important couriers, Klein and Drach, was a heavy blow for the Westerweel group. Because they were half-Jewish and did not have to wear a star, they could move more freely and were not subjected to all sorts of limitations. Their tasks were taken over as well as possible by several other helpers. In the first place, this was the unshakeable and resourceful Lore Durlacher, who had false papers.

The orthodox Harry Asscher also became a helper. For this purpose, Harry travelled throughout the country with false papers as Hendrik Zijlstra, an employee of the Dutch PTT company, and had 'actually no fixed address for two years'.

Other members of the group, such as Wil Westerweel, Menachem Pinkhof, Mirjam Waterman, Jan Smit and Bouke Koning, were also busy caring for pioneers in hiding in the autumn of 1943. Lore Sieskind stayed in hiding in Haarlem for a few weeks. However, she then went to live in rooms in The Hague with false

---

4 GFHA, Cat. no. 70, Verklaring Gideon Drach, juli 1959, with supplement 1 december 1987.

papers. Here, she worked with Joop Westerweel with caring for and transporting pioneers in hiding.[5]

At the beginning of October 1943, the Westerweel group was again hit by arrests. Almost from the beginning of resistance work, the group had a *safe house* in Rotterdam. First this was the house of Tine Segboers brother in the Van Nideckstraat in the Blijdorp neighborhood. In this house, which was at walking distance from the Central Station, pioneers who had to leave their hiding place for whatever reason could be offered sanctuary. Shushu Simon used the house as a base camp for his journeys to France.

After the arrest of Simon in January 1943, the group cleared this house and set up a new refuge in the Dordtsestraatweg in the southern part of Rotterdam. This home belonged to a former non-Jewish pupil of the *Werkplaats*, an acquaintance of the Westerweels, who left for Belgium or France at the beginning of 1943 to avoid the *Arbeitseinsatz* in Germany. Joop Westerweel had helped him obtain false papers. To show his gratitude, he had given Joop his identification card and the management over his house. With this identification card, a Palestine pioneer had then gone to Germany in his place.[6]

The manager of the shelter was the 28-year-old Letty Rudelsheim. She came from an Amsterdam Zionist orthodox family. After an education as a teacher, she went in 1938 to the religious *hachshara* in Beverwijk. When the Germans put a stop to this course at the beginning of the occupation, Letty went to the *Jeugdalijah* house in Loosdrecht.

After she went into hiding with the Loosdrecht group in August 1942, she stayed at various addresses in the provinces of North Holland and Friesland. At the beginning of 1943, Letty went to Rotterdam, where she helped Wil Westerweel in the household. Wil was very busy after the birth of her son Bart, several months before. In the spring of 1943, Letty moved to the Dordtsestraatweg, where she opened a practice as a speech therapist, as she told the neighbors and shopkeepers nearby. Letty did not look Jewish and had a good forged identification card in the name of Alida Jonker. The speech therapy practice was also an explanation for the large number of visitors at her home.

There was strict security with the new shelter, whereby 'a secret manner of ringing the bell' was agreed, so that Letty could recognize 'our' people. To hide the high water consumption, she bought extra hot water from a laundry. 'I had to

---

5 GFHA, Cat. no. 9, Verklaring zonder datum Harry Asscher, waarschijnlijk 1955, 1–3. See also GFHA, Cat. no. 217, Verklaring Lore Sieskind, 5. Westerweel, Lijn of cirkel, 69.
6 Westerweel, Lijn of cirkel, 70. In other sources, a house is mentioned in the Strevelsweg. This street was more or less an extension of the Dordtsestraatweg.

take account of everything: I used the food ration cards at different shops and always bought small amounts to avoid suspicion.'[7]

Initially, everything went well—although there were several alarms. One day, for example, a woman rang the doorbell and said that she needed a place to hide. Letty and the three people present fled immediately to Bouke Koning in Groenekan, near Utrecht, where they stayed for some time. In the meantime, several members of the Rotterdam resistance kept an eye on the house in the Dordtsestraatweg. When there did not seem to be any immediate danger, Letty returned. On another occasion she was summoned to appear at the town hall for an investigation into the disappearance of the owner of the house. Letty said that he had gone to Germany; she was told that this would be investigated further.

At 11 o'clock in the evening of 10 October 1943, the bell was rung with the 'special ring'. When Letty opened the door, there were three *SiPo* agents. 'They screamed: "Stand still!" Petrified, I stood in the door opening, waiting.' All eight people in hiding in the house were arrested. These included Kurt Hannemann, who had temporarily moved to Rotterdam in connection with the arrival of Norbert Klein in the Gerritsen's house, and Rachel Cohen, the daughter of Ru Cohen, of the Deventer Union.[8]

They were taken to the police headquarters in Rotterdam. The next day, the eight people were first transported to Camp Vught and, several days later, as 'punishment cases' by train to Westerbork. Four of them, Hannemann, Manfred Rübner and Harald Simon and his wife Meta were handcuffed and transported separately from the others and guarded by Dutch policemen. In Westerbork, the *hachshara* group, which was warned that Hannemann was on his way, had an escape plan ready. However, with the heavy guard on Hannemann, it proved impossible. 'It was the first time that we experienced such a display and it proved how important Kurt was in their eyes', Paul Siegel realised.[9] Kurt Hannemann and the three others were immediately taken to the waiting train for Auschwitz. The other four went with the next train. None of the eight would return.

---

[7] Letty Ben Heled-Rudelsheim and Miriam Dubi-Gazan, *Gesprekken met mijzelf in Auschwitz* (Kampen: Kok, 2003), mainly 32–41, 44–48, 61–69.
[8] Ben Heled-Rudelsheim and Dubi-Gazan, 73–78. See also Interview Sytske De Jong-Letty Rudelsheim, 22 november 1998, 11.
[9] Siegel, *Locomotieven*, 124, 125.

Fig. 24: Letty Rudelsheim, sitting far right, was the manager of the safe house in Rotterdam. The photograph was taken during a meeting of pioneers in Deventer at the beginning of the war.

A Trap Set Up around Letty Rudelsheim

Because of her false papers, Letty Rudelsheim, who was locked up in the Scheveningen prison, was treated by the Germans as a non-Jewish woman who had helped Jews. Probably after betrayal by a spy in her cell, she had to admit that she was Jewish. Afterwards, the Germans used her under her false identity of Alida Jonker as a sort of bait for the Westerweel group.

At the end of November 1943, Karel Kaufmann, a Jewish helper of the group (more about him later), told Chiel Salomé that it was possible to buy the freedom of 'Alida Jonker', who had in the meantime been sentenced to a six-month prison term for helping Jews. A prison guard from Scheveningen could be bribed to take Letty/Alida to Rotterdam and release her at the station for a large sum of money.

At the beginning of December, Chiel told this to the others in the weekly meeting of the Westerweel group in Rotterdam. Wil Westerweel immediately found it a 'scary business, because if someone only wants to save a person's life for lots of money, he could also be ready to betray us for lots of money as well'. However, a majority of the people present supported the liberation attempt, at which she decided to cooperate.[10]

After a short time, Letty, who had been transferred to Camp Vught for the rest of her prison term, was taken back to The Hague by the Germans. Here, she was given a letter in Joop Westerweel's handwriting stating, 'I am ready to pay ransom for the release of Alida Jonker in the amount of three thousand guilders'. One of the German agents told her that they had agreed with this proposal. Letty realized that a trap had been set for the Westerweel group's members. However, she could do little else than cooperate.

On 15 December, the money for the ransom had all been collected. It was agreed

> 'that Chiel and I (Wil Westerweel) would go to the station this Wednesday afternoon around 2 o'clock. As soon as Letty and her attendant had gotten off the train, Chiel would take the man to the main exit with him to get the money from the bank together. I would take Letty to the rear exit, where I had two bicycles standing ready. When I saw Letty, I stood inconspicuously next to her and whispered: "You don't know me; walk behind me." Chiel disappeared with the man'.[11]

That was the last thing that went according to plan. When Wil and Letty got to the rear of the station, six German agents awaited them. They were arrested, as was Chiel Salomé. There were no more arrests, however, because Harry Asscher happened to witness the arrest of Wil and Letty, on his way to the Westerweel family's house at the Provenierssingel, which was located immediately behind the station. He warned Thea Mok, the Westerweel's Jewish housekeeper. Together, they fetched illegal papers from a hiding place and dressed the Westerweel's son Bartje, who was taken to acquaintances. Thea then quickly went to the Montessori school on the Beukelsdijk to warn Joop that he should immediately go into hiding.[12]

Betrayal by an Infiltrator

It is not completely clear how the betrayal of Westerweel group in Rotterdam occurred. The Germans would have formed a picture of the group's operation from various interrogations. Other matters, such as the disappearance of the original

---

10 GFHA, Cat. no. 201, Interview Chiel Salomé probably by Mirjam Pinkhof-Waterman, 22 april 1986. See also Westerweel, Lijn of cirkel, 75.
11 Westerweel, ibid., 75, 76. Ben Heled-Rudelsheim and Dubi-Gazan, *Gesprekken*, 84, 85.
12 Westerweel, ibid., 76.

inhabitant of the house in the Dordtsestraatweg, could also have played a role. As described in the stories of the two groups of five pioneers who were in hiding in Bielefeld and Dortmund through the *Arbeitseinsatz*, the police in Germany checked the identity of the Dutch workers who went there to work. It is quite possible that the police found something irregular about the papers of the pioneer who had taken the place of the man from the Dordtsestraatweg. Nothing is known about his fate.

The investigations into the original inhabitant could have focused the police's attention on 'Alida Jonker', the speech therapist who worked there and received so many visitors. Furthermore, there was an infiltrator in the Westerweel group who played a part in the betrayal.

According to various sources, this was the previously-mentioned Jewish student Karel Kaufmann. From the beginning of 1943, he was active as a helper of Kurt Reilinger and Dirk van Schaik in Gouda and Rotterdam, where he also lived. Karel Kaufmann was born in Hengelo in June 1920. His parents came from the east of the country and had settled in Rotterdam in the mid-1920s. However, the family, which then consisted of five people with two younger sons next to Karel, had gone to Nijmegen in 1934. Here, he also attended High school. In September 1938, he began his study of economics at the then Katholieke Hogeschool (Catholic Academy) in Tilburg. A year later, he continued his study at the Economische Hogeschool (School of Economics) in Rotterdam.[13] This was no longer possible from 15 September 1942, due to anti-Jewish measures.

Karel Kaufmann's official address in Rotterdam was the G.W. Burgerplein, where he probably rented rooms.[14] The Burgerplein was a smart location with large mansions in the borough of Middelland. In this neighborhood, close to the Burgerplein, was also the Beukelsdijk, the location of Joop Westerweel's Montessori school. It is possible that the two of them met each other here. According to Menachem Pinkhof, Kaufmann had joined the group through Westerweel.

Karel Kaufmann, or Piet, as he called himself in the resistance, was involved in the housing of several pioneers in Rotterdam and their transport to the Belgian border. These included Berrie Asscher and Hans Mogendorff. Mogendorff's wife, Fieke (Sophie) de Winter came from Nijmegen and knew Kaufmann from the time that he lived there.

Hans Mogendorff and his wife travelled with him to the Belgian border near Putte in June or July 1943. Kaufmann introduced them to the guide and warned

---

13 Stads Archief Rotterdam, Informatie gezin Elie Kaufmann.
14 https//www.joodsmonument.nl., accessed 14 augustus 2014, Informatie over Carel Albert Kaufmann.

Hans not to pay until they were in Belgium; he left them there. The Mogendorffs were part of a large group of more than 30 people that was led to the border by the guide. When they arrived, the guide told them that he had to go to a lookout post to be certain that everything was safe. However, a short time later, German soldiers appeared, surrounded the group and arrested everyone.[15]

Hans Mogendorff was convinced, albeit without much evidence, that Karel Kaufmann had betrayed them. Because he and his wife stuck to their non-Jewish identity during the interrogation, he was not sent as a 'punishment case' to Westerbork, but to the *Arbeitseinsatz* in Essen. At the end of July or the beginning of August 1943, when he got leave after a bombardment, Mogendorff travelled back specially to Amsterdam to warn Kurt Reilinger and Kurt Hannemann to avoid 'any further contact' with the 'traitor' Karel Kaufmann.[16] Reilinger went to France a short time later; it is not clear what was done with Mogendorff's warning.

It is not known when Kaufmann began working with the Germans. According to a warning in the *Signalementenblad*, published by the resistance, this happened after his arrest and transfer to Westerbork (the *Signalementenblad*, first published in 1943, contained a list of presumed collaborators, with some photographs, names, aliases and descriptions). Kaufmann might have taken this step to avoid the arrest of his family members. This seems to correspond with other information. His parents and two younger brothers John (born 1924) and Martijn (born 1926), who had gone into hiding from Nijmegen from the autumn of 1942, remained free until February 1944.

Karel Kaufmann had explained to others that it was his intention to mislead the Germans—but little came of this. Kaufmann, who according to the *Signalementenblad* did not look Jewish and also used the name K. van Smalen, was described as 'very dangerous' and as one who had made 'many victims'.[17]

The Consequences of the Affair

After her arrest, Letty Rudelsheim was taken back to the prison in Scheveningen, where she was interrogated several times. However, apart from her work as manager of the safe house, she had not been active in the Westerweel group and had little to tell the Germans. She shifted the guilt as much as possible to Shushu

---

15 Mageen, *Van zonsondergang*, 30, 31. See also Asscher, *Van Mokum*, 121–126.
16 Mageen, *ibid.*, 40.
17 NIOD, *SIgnalementenblad no. 3*, augustus 1944, no. 530, Karel Kaufmann.

Simon, who was dead. Apparently forgotten, she languished another six weeks in Scheveningen.

At the end of January 1944, Letty was taken to Westerbork as a 'punishment case'. The car that transported her from the prison to the Hague station was driven by a man who the Germans called the 'personal chauffeur' of Franz Fischer, the infamous deputy head of department IV-B4 in The Hague, who was tasked with finding Jews who were in hiding.

At Westerbork, Letty was reunited with the other members of the Rudelsheim family. In the meantime, her brother had been able to establish that they all had double British-Dutch nationality through their father, who was born in London. For this reason, Letty was able to stay in Westerbork for some time. On 25 March 1944, however, she was deported to Auschwitz. Because of her British citizenship, she was put in a separate wagon with 38 'professors and artists' and others, who were half-Jewish or, like Letty, had a comparable status.

One of her fellow passengers in the wagon was the previously-mentioned 'chauffeur of Fischer'. This was almost certainly Karel Kaufmann. He had gone into hiding in Gouda and was arrested on 16 February. His parents and two brothers had been arrested the previous day from their hiding address in Rotterdam.

Karel Kaufmann was imprisoned in the Rotterdam police station for three weeks. Here, he was kept strictly separated from the other prisoners and his family. On 8 March, he was taken to Westerbork as a 'punishment case' and from there to Auschwitz on 25 March with the same transport as Letty Rudelsheim. He probably thanked his privileged status as half-Jew to his work as an employee of the German police. 'There is a spy among us', Letty warned the other passengers in the wagon.[18] His status did not help Kaufmann. He was put to work in the Buna-Monowitz complex. However, after several days he was hospitalized, and was returned to Auschwitz, where he died at the beginning of April 1944.

His parents and brothers, whom he had wanted to protect, arrived in Westerbork on 1 March 1944. Two days later, they were all transported to Auschwitz. His parents were murdered immediately; his two brothers died in 'Central Europe' at the end of July 1944.[19]

---

18 Ben Heled-Rudelsheim and Dubi-Gazan, *Gesprekken*, 88–92. Letty Rudelsheim describes Fischer as the head of the Gestapo in the Netherlands with headquarters in The Hague. She almost certainly means the SS man Franz Fischer, the head of the Referat IV-B4 (Office of Jewish Affairs).
19 Stads Archief Rotterdam, Arrestantenkaarten, Jong tot en met Klaassen, 1944, no. 3704 and 3550 no. 8 A, familie Kaufmann. A bonus of 7.50 guilders was paid out for all five arrests. See also Archief Herinneringscentrum Kamp Westerbork, gezin Elie Kaufmann.

Upon her arrival in Auschwitz, Letty evaded the infamous selection. She was immediately taken to the camp with her possessions. This brought an end to her privileged position. She was forced to participate with the other prisoners in the gruelling camp existence; with the help of several fellow prisoners, Letty (a slender woman) managed to survive. She also survived the death marches when Auschwitz was cleared out. She was liberated in April 1945 in the Neustad Glewe camp in Eastern Germany, weighing only 28 kilograms. After many years of convalescence in Switzerland because of a serious case of tuberculosis, she was able to immigrate to Israel in 1950.[20]

Wil Westerweel and Chiel Salomé were incarcerated in the Rotterdam police headquarters on the Haagsche Veer and interrogated at the SD headquarters on the Heemraadsingel. The Germans regarded them as people who helped Jews; Chiel had been imprisoned earlier for this help. The Germans now knew that Wil was a key person in her husband's organization. She was interrogated by men such as the infamous SD man Hoffman, who treated her fairly reasonably. He was mainly interested in where the rest of the family members were hiding. But there was nothing that Wil knew or wanted to tell. On New Year's Day 1944, she received a postcard with 'Happy New Year' on it in Joop's handwriting, which reassured and encouraged her.

Several weeks later, she was transferred to Camp Vught. After the camp was cleared out at the beginning of September 1944, Wil was transported to the infamous women's internment camp Ravensbrück, about 80 kilometers north of Berlin. Insufficient food and clothing, 11-hour work days and later also overcrowding and contagious diseases made the circumstances in the camp inhumanly heavy. Around 20 April 1945, the Red Cross took the Dutch and Belgian women from Ravensbrück to Sweden, where they could recuperate in freedom. The sick and emaciated Wil regained her strength there and began to adjust to a normal life. She returned to the Netherlands in the summer of 1945.[21]

Chiel Salomé was imprisoned for six weeks at the Rotterdam police headquarters. Initially, the Germans interrogated him almost every day, but did not mistreat him. He told them nothing, but Chiel did tell several Dutch policemen about his help to Jews. At the end of December 1943, he was transported to Camp Vught; six weeks later, he was sent to the Dachau concentration camp near Munich. In Dachau, he did mainly factory work, but he was also in the camp prison for some time. After this, he became barracks head—'not the worst job', he said later. The Americans liberated Dachau at the end of April

---

20 Interview Sytske De Jong-Letty Rudelsheim, 22 november, 1998.
21 Westerweel, Lijn of cirkel, 76–96.

1945. Sometime afterwards, Chiel travelled back to the Netherlands with a group of Dutch prisoners of war.[22]

The safe house affair in the Dordtsestraatweg also led to the arrest of Lore Sieskind. Around 12 October 1943, Lore had agreed to meet Dirk Kroon, a resistance man with whom the Westerweel group worked incidentally and who had been approached to check if there was a possibility of liberating Kurt Hannemann (who had been arrested in Rotterdam). However, Dirk arrived much too late for his appointment with Lore at the Haarlem train station. While she was waiting, she was recognized by a member of the Amsterdam SiPo. She was locked up in the Haarlem police station, interrogated and mistreated by continually being slapped on her face. The Germans offered her 'life and freedom' if she would cooperate with them. However, she refused.

After five days, Lore was (on the basis of a forged marriage certificate to a non-Jewish man, which was sent to her shortly before the war by her parents who had fled to Bolivia) transferred to the Hollandse Schouwburg in Amsterdam. Several attempts to liberate her by Frans Gerritsen and Menachem Pinkhof failed. Finally, Frans negotiated successfully with Walter Süsskind, the Jewish manager of the Schouwburg, to obtain her release. By this time Lore was too well-known and could be of little use for the organization. At the beginning of November 1943, Joop Westerweel therefore took her to Brussels. From there, she went on to Paris.[23]

Consequences for the Organization

If the Dordtsestraatweg arrests had occurred several months earlier, this might have been a serious matter for the Westerweel group. By the end of 1943, however, a large number of Palestine pioneers had already been taken to France. Moreover, because Wil, Chiel and Lore had told the Germans nothing important, the work could basically continue. In January 1944, Joop Westerweel was able to establish a new refuge for escapees from Westerbork in Pim and Gerda van Maanen's home in The Hague. The work of what could be called the 'Haarlem/ Amsterdam branch', around Frans Gerritsen and Jan Smit, Lore Durlacher and Tinus Schabbing, went on.

After his wife's arrest, Joop Westerweel went into hiding with acquaintances in Rotterdam. He first stayed with his colleague Greet Kolff of the Montessori

---

22 GFHA, Cat. no. 201, Interview Mirjam Pinkhof-Waterman-Chiel Salomé, 22 april 1986.
23 GFHA, Cat. no. 217, Verklaring Lore Sieskind, 6, 7.

school. She was amazed at the risks that Joop took. Because the amount of work he did left little time to go shopping and cook, he persuaded her to go to Martin's, a restaurant in the Rotterdam center, for dinner. The doorman could 'be thoroughly trusted'. She had serious doubts but could not resist his 'compelling conviction'. They had hardly sat down in the restaurant when the police, possibly warned by the doorman, appeared and arrested both of them. Joop was only able to escape through the clumsiness of a police officer, who let him go in front of him through the revolving door. Joop pushed the door back with all his might, ran away and eventually rang the doorbell at a random house some distance away. Here, he asked if he could come in for a while. The people in the house let him in and he evaded arrest.

Joop Westerweel realized that he was being sought and that it was highly likely that his description was circulated. He therefore disguised himself by having his hair cut short, growing a small moustache and wearing spectacles. Yet, he also attended the Christmas celebrations at the Montessori school at the end of December 1943. Under his leadership, the children performed the play that he had written from the book *Het Kerstekind* (The Christmas Child) by Stijn Streuvels. 'The children were all in on the conspiracy. The school gate was strictly guarded, and an escape route was made for Joop. The Christmas play was performed in this atmosphere; I can quite easily say in an atmosphere of high tension and super-terrestrial haste.' Immediately after the performance, he disappeared again.[24]

This was once again a risky undertaking; his colleague Greet had the impression that Joop (who used the illegal name Victor) was no longer himself. His short hair symbolized this, to some extent. Joop also seemed to perceive this, and referred to the Old Testament hero Samson, who lost his strength after his hair was cut off. The man who was previously able to encourage and inspire everyone during setbacks, now made a dejected impression on other members of the group. His wife was in prison and his children had gone into hiding. He was now forced to realize that he no longer had any influence on their lives. Harry Asscher, who cared for Joop's daughter Ruth in Friesland, told Joop during a meeting that the people who hid them had cut off her braids. 'Joop reacted sadly to this, and said softly: But I thought that I was still alive.'[25]

It would perhaps have been sensible to allow Joop Westerweel some time to rest and to give him a break from his participation in resistance work. However,

---

24 Greet Canter Visscher-Kolff, "Herinneringen van een Montessorileidster," Westerweel, *Verzet*, 61, 62.
25 Naftali (Harry) Asscher, "Naftali Asscher," Westerweel, *Verzet*, 88.

this did not happen—he would have undoubtedly immediately rejected any such idea. In fact, Joop did exactly the opposite of resting. He tried to regain a grip on his life by fully devoting himself to the resistance. Because he was in hiding and no longer had to work, he now had more time than before.

He took pioneers across the border near Budel several times and then by train to Brussels via Hamont. In Paris, he had discussions with Max Windmüller and other prominent chalutzim who were staying in France. At the end of February 1944, he appeared at a farmer's barn at the foot of the Mont Valier to the south of St. Girons in the Pyrenees. Here, twenty pioneers and others (in total 34 people) were waiting to go across the mountains to Spain. Joop's dangerous journey to the Mont Valier—the region was forbidden for non-residents—was organized by Lore Sieskind as a sort of tribute to him from the *Hechalutz*.

The pioneers from the group collected together in a field outside the barn and listened to Joop's short farewell speech. According to Paul Siegel, Joop spoke calmly, straight from the heart and without becoming maudlin. He wished them a successful crossing to Spain and expressed the hope that they would succeed in reaching Palestine. A new state had to be built there, in which Jews from the whole world could go and live. He urged those present never to forget their comrades who had sacrificed their lives in the struggle for their liberation.[26] His words made a huge impression on them.

The Arrest of Joop Westerweel and Bouke Koning

Some of the people present that day at the foot of the Pyrenees had the impression that Joop Westerweel was tired and lonely. This could obviously be attributed to Wil's arrest and the consequences for his family—but also the changed circumstances in the group (whereby the focus of the activities had shifted to France) played their part. While he had at one time felt he was the informal leader, he now had the feeling that he had become someone who carried out only the occasional odd job.

In the Netherlands, some people felt that he shut himself off from criticism and took irresponsible risks—for example, by going to the Spanish border and by engaging in extramarital affairs. On 2 March 1944, Frans Gerritsen wrote to Lore Sieskind in France: 'Recently, they have done Guusje (Joop Westerweel) a very great injustice and said things that caused him great pain'. In Brussels, Paul Sie-

---

26 Siegel, *Locomotieven*, 155. There are different versions of Joop Westerweel's speech, which are, however, all about the same.

gel heard a 'heated argument between Tinus Schabbing and Joop, whereby Tinus mentioned the great dangers of the border crossing at Budel. The two parted without resolving their conflict.[27]

At the beginning of March 1944, the subject of transporting two pioneers to Belgium came up during a meeting of the Westerweel group in Amsterdam. Because everyone was busy, Joop Westerweel offered to take them. Perhaps because of the criticism on the crossing at Budel, he asked Bouke Koning to accompany him. Bouke had previously taken several pioneers to Budel, but he had always done this alone—and he had also never crossed the border.

Bouke Koning agreed reluctantly. He was in fact just as ill-suited as Joop for resistance work. He had only been released from prison shortly before the meeting in Amsterdam. He had been arrested in mid-December 1943 for supplying ration cards to a resistance group of Utrecht students (resistance activity that was separate from the Westerweel group). One of the students was arrested and had mentioned Bouke's name during interrogation, after which the Germans raided his house in Groenekan. During the ensuing search, no incriminating items were found, but they did find Lotte Wahrhaftig, who was Jewish. After first being imprisoned in the police station in De Bilt for several weeks, Bouke was taken to the *Huis van Bewaring* in Amsterdam to serve a sentence of six weeks. This was the usual price for helping Jews.

On 10 March 1944, he and Joop set forth to transport two Loosdrecht Palestine pioneers—the friends Ruth Direktor and Thea Perlmutter, 18 and 19 years old, respectively—to Belgium. After staying the night in Eindhoven, they spoke together in the train about the fact that they were really too vulnerable to do this work. Bouke had to wear a hat, because his hair had been shaved off in the prison. Joop was sought after by the Germans. Moreover, the fact that there were two guides to take two girls away was suspicious. 'Joop, what have we actually started?', Bouke asked him in the train. But the question fell upon deaf ears with his travelling companion. 'Yes, Joop was a go-getter and a bit of an adventurer—and nothing would stop him', Bouke explained after the war.[28]

Things went wrong in Budel. On the long quiet road to the border, they were passed by two Landwachters (NSB auxiliary policemen) on bicycles, who probably warned a German guard post further on. When Joop went ahead to reconnoiter the road, he walked into the trap and was arrested. Bouke waited with

---

27 GFHA, Cat. no. 63, Brief Frans Gerritsen aan Teuntje Kwak (Lore Sieskind), 2 maart 1944. See also Siegel, *Locomotieven*, 148.
28 GFHA, Cat. no. 117, Interview Yigal (Kurt) Benjamin-Bouke Koning, 9 september 1989, 4, 5.

Ruth and Thea at a trusted café. After some time, a German patrol arrived and arrested them.

After this, several other things went awry, partly because they had not thought up a good alibi beforehand. For example, the girls confessed almost immediately that they were Jewish, which made the usual story of buying tobacco in Belgium implausible. In addition, Joop Westerweel used an identity document from the Antwerp smuggler Henri Lelièvre, which he had received from him earlier. However, despite Lelièvre's German imprisonment, he appeared to be in a search register. The Germans also found a letter with information about the resistance. Bouke had immediately said to Ruth Direktor: 'get rid of it', but her attempt to flush the letter down the toilet failed.

Ruth and Thea were locked up in the Dutch MP barracks in Budel. Bouke and Joop were imprisoned in a cell under the town hall, guarded by an older village policeman. They tried to get him to release them, but he did not dare to. Later, Bouke realized that they should have overpowered the man at night or should have offered him money, after which they could have fled. The next day, the Germans took all four of them to police headquarters in Eindhoven.[29]

Escape Plans

The Westerweel group was soon aware of the arrest of Joop and Bouke. The Budel pub landlady called an agreed alarm number in Amsterdam, which is how Jan Smit heard the news. He travelled immediately to Eindhoven via Budel to speak to the woman. Here, as a result of helper Van Heeckeren, who had many contacts in the city, he was able after several days to make an appointment with a 'good' policeman. This policeman took him to the station, where Jan spoke to with Joop Westerweel for five minutes. Joop gave Jan the impression that he had recovered from the shock of being arrested, 'Victor had become Joop again', and had an escape plan ready.

Joop had heard that he would be taken to Rotterdam, where the SD had a file on him. During the journey from Eindhoven to Rotterdam, he wanted to jump from the train through the toilet window, after which a car would pick him up. But nothing came of this plan. Joop was taken unexpectedly and under heavy guard to the SD headquarters at the Heemraadsingel in Rotterdam. Moreover, his guards had shoved a stick down his trouser leg to prevent him from escaping.[30]

---

[29] Ibid. See also interview Mirjam Pinkhof Waterman-Bouke Koning, 20 april 1990, 2–4.
[30] GFHA, Cat. no. 225, Brief Jan Smit aan Mirjam Pinkhof-Waterman, 25 mei 1964.

In Eindhoven, Joop Westerweel still had high hopes. In several letters smuggled to the outside, he encouraged his comrades and gave instructions to avoid further arrests. He thought that his case could result in a 'stay in Vught'. However, the gravity of his situation soon became apparent in Rotterdam, where they produced a file on him, in which was stated that he was a communist who had killed a policeman and had committed acts of sabotage, among other things. Joop thought that the file was mainly based on statements by Karel Kaufmann, who had given his imagination free rein.

It is rather improbable that the Germans themselves believed the accusations. However, they did offer them the opportunity to put pressure on Joop Westerweel to show 'with absolute certainty that matters were different'. This pressure to force him to reveal real facts amounted to heavy mistreatment. The Germans forced him to remain standing, handcuffed for almost two days, from two o'clock Thursday afternoon until eleven o'clock Saturday morning, with several short breaks to eat and drink something. In between, he was interrogated 'in a dark basement cell' by a policeman and was for 'a time bent and handcuffed with all my clothes pulled over my head'. At each answer 'I could expect a kick or a lash of a whip'.

On Sunday morning, he was told that he was allowed to write a farewell letter to Wil because he would be taken to the *Standgericht* (summary court-martial), which usually pronounced death sentences. When he was writing the farewell letter, he was interrupted and subjected to further interrogation. On Sunday afternoon he got some rest, but on Monday the interrogations would resume.[31]

Numbed by the torture he had endured, Joop decided that Sunday 'to write a story, in which I admit as much as I can, but in which I moreover do not provide any single name'. He understood, however, that heavy pressure would then follow to give names of other members of the group. He asked therefore in a letter that he managed to get smuggled outside for permission to:

> 'betray you, in other words: I would like to say: "OK then", and give them names and addresses. However, these would be addresses or telephone numbers that do not reveal that I've tipped you off, because it would then be naturally much worse. When the police arrive at all these addresses, the people there must say, "Sure, he/she lives here, but he/she has not appeared for the last few days". This will create the impression that the list was good, but they've all gone into hiding.'

---

[31] Joop Westerweel, "Brieven van Joop uit de gevangenis," Westerweel, *Verzet*, 47–50. The letters are undated and written to 'Els (Lotte Durlacher), beste vrienden, beste mensen.'

He then listed the names and addresses of people who he wanted to mention in his 'confession' and who had to be warned.[32]

Joop's letter reached Frans Gerritsen or Jan Smit that Sunday evening, through a 'good' employee of the police headquarters at the Haagse Veer. They immediately organized a late-night meeting of the group in Vreeland near Loosdrecht and wrote a reply. On Monday morning at five o'clock, each travelled his own way to Rotterdam to give their reply to Henk Brusse, the man who organized the contact with Joop. Brusse was an architect and a relation of Jan Smit, who was active in the Rotterdam resistance.

Henk Brusse also had a plan ready to liberate Joop. His general practitioner took blood from Frans and Jan, from which powders could made that simulated the effects of gastric bleeding. When Joop Westerweel was taken to hospital, an armed resistance group that Brusse was in contact with would liberate him.

However, this plan also failed. Joop took the powders and they worked, but the Germans transported him to the Zuider Hospital in Rotterdam, where he was heavily guarded, partly because another prisoner was liberated a week earlier. This meant that the armed resistance group could not reach him. Later, Joop was transferred to the hospital of Camp Vught.[33]

During this very difficult time, Joop Westerweel had great support from his partially regained faith. It was precisely just after the heaviest interrogation that he felt as if he was being crucified 'in the full revealing light'. 'A human being is only put on that cross once. This is how I was the last time I wrote to you; what a wideness, a supra-temporal mercy. It stayed that way for the first days; Sunday, Monday and Tuesday were so happy. There was actually no cell at all, only space, wideness.'

He also greatly appreciated the contact with the other members of the group, who had all added something to the letter that Jan and Frans had written and taken to Rotterdam. 'I felt as if I was completely with you. It made me feel indescribably good. We have come to mean a great deal to each other in the course of this time', he wrote from the camp hospital in Vught.

The uncertainty was the biggest problem for Joop. He often put his hands together and prayed: 'Father let this goblet pass me by'. Surrendering himself to God's will was not yet possible, as he was haunted by the errors and mistakes that he had possibly made. He regarded himself as responsible for the arrest in Budel; he should perhaps have jumped out of the train between Eindhoven and

---

[32] Westerweel, "Brieven," 51–53.
[33] GFHA, Cat. no. 225, Brief Jan Smit aan Mirjam Pinkhof-Waterman, 25 mei 1964.

Rotterdam. He was also not satisfied with the 'stupid statements' that he had provided to the Germans.

He wanted to escape alone so that he would not bring his friends into danger. But he also had wife and children and friends—and wanted to live. 'That's how it goes. I am just a normal person: you may never idealize me. I have to get through these bitter days.' Should it go wrong, then he would go as a man. 'That is not the most difficult part, but what lies between: The road from Gethsemane to Golgotha.' He left it up to his friends whether they wanted to make another liberation attempt, but he also made it known that he dreaded further interrogation.[34]

Joop Westerweel's letters from Camp Vught were smuggled outside by Dr. Steyns, a gynaecologist from Utrecht. He went to the camp at set times to treat prostitutes who had venereal diseases, and were imprisoned there because of the danger of infection for German soldiers. It is a pity that the answers to Joop's letters were not kept. It is certain that not all members had the same opinion on Joop's position and about what should be done. Menachem Pinkhof, for example, thought that Joop saw his position as too somber. He was in no danger of getting the death sentence, in Menachems opinion.[35]

Another idea in the attempts to improve Joop's position was to take formal legal action. People considered bringing charges against Karel Kaufmann because of his incorrect statements. Two people, one of whom was a doctor, stated that 'Mr. Westerweel' had taken German wounded to a Wassenaar hospital 'even with danger for his own life' in May 1940.[36] These attempts, however, apparently did not produce any results. A majority of the group's members was therefore in favor of continuing the liberation attempts, mainly because Joop would otherwise try to escape on his own, with all the associated risks.

Joop's next letter listed five possibilities to escape. One of these possibilities could be by cutting through the barbed wire and running away. However, it was better to use the papers of one of the Philips employees who regularly went to camp Vught to visit the so-called *Philips Kommando*.

Joop wrote a letter, probably at the end of March 1944, in which he reported that he was allowed to stay in the hospital barracks of Vught for another three

---

34 Westerweel, "Brieven," 53–56.
35 GFHA, Cat. no. 173, Brief Menachem Pinkhof, smuggled out of Westerbork, at the beginning of May 1944. The passage with the remark about Joop Westerweel was later crossed out, but can still be clearly read.
36 Collectie B. Westerweel, Verklaringen P. Boef and C.M. Versteeg-Solleveld, 4 en 6 april 1944.

weeks. He wondered, however, if he 'as a moreover healthy person' could find enough patience to endure it.³⁷

A New German Trap

Shortly afterwards, the matter of the escape gained momentum. Joop Westerweel had sent his first letters that were smuggled outside to Jan Smit in Amsterdam. However, Jan did not have a telephone. Therefore, Joop sent his later letters, which were legal, as part of the correspondence that went through the censor to Gerda van Maanen. Gerda did have a telephone, which made it easy for the contacts with the other members of the group. Therefore, Gerda and Pim's address was known to the Germans.

The Germans may have become alarmed by the censor because of several cryptic passages in the letters, which made them suspect coded messages. In any case, there followed a thorough check of Dr. Steyns, in which they found a letter from Joop about his escape plans. At the beginning of April, the Germans decided again to use an infiltrator, who would try to prevent an escape by Joop Westerweel and eliminate the rest of the group.

Not much is known about this infiltrator. He used the first name 'Karel', but it is not clear if this was his real name. It is also not clear to what extent he deliberately cooperated with the Germans. 'Karel' was almost certainly a former prisoner of Vught who Joop Westerweel had met there. 'Karel' may have led the *SiPo/SD* to the Westerweel group unconsciously. After his arrest, he gave the Germans the information they wanted about the group, possibly under the pressure. After the war, he was sentenced to a relatively light sentence because it could not be proved that he had betrayed people on purpose.³⁸ The use of the name 'Karel' led to the misunderstanding that he could be the same person as Karel Kaufmann. This is not the case.

'Karel' reported to Pim and Gerda van Maanen in mid-April 1944. He had a letter in Joop Westerweel's handwriting that stated: 'You can trust Karel; give him all your cooperation'. He also said that Joop's letters had been intercepted and that the Germans were now aware of the escape plans. A new plan had to

---

37 Westerweel, "Brieven," 59, 60. See also GFHA, Cat. no. 173, Brief Menachem Pinkhof smuggled out of Westerbork at the beginning of May 1944.
38 GFHA, Cat. no. 63, Brief Frans Gerritsen aan Mirjam Pinkhof, 7 april 1964. See also GFHA, Cat. no. 252, draft text of Interview Mirjam Pinkhof-Ellen Landweber, 14, 1988. GFHA, Cat. no. 173, Verklaringen Pinkhof.

be made and he wanted to help. He knew Camp Vught well and could restore the connection with Joop.

The Van Maanens, who had little experience in resistance work apart from making forgeries, were impressed and gave 'Karel' a room in their home. He stayed there with pioneer Paul 'Amo' Rosenbaum, who had escaped from Westerbork shortly before and who told him a lot about the resistance work in the camp.[39]

Together with 'Karel', Jan Smit, Lore Durlacher, Menachem Pinkhof, Mirjam Waterman and Pim van Maanen and several others began to discuss ways to liberate Joop from Vught. The discussions were held in the house of Tine Segboer, a friend of Wil Westerweel, who lived in The Hague. However, several people had doubts about the reliability of messenger 'Karel', about whom nothing further was known. For example, Menachem Pinkhof thought that it was suspicious that he had stayed at the Van Maanen's house for some time. Mirjam Waterman had grave doubts about 'Karel' and his escape plans. For them, the choice was to continue the operation despite the uncertainties or to stop it. Out of solidarity with Joop and the others, they chose to continue the contacts.

Pim van Maanen had also begun to doubt. He had received a note from Joop in Vught, which read: 'Pim and Gerda must leave immediately.' They then moved quickly to another address. Pim then showed the note to 'Karel', who did not know of the move. 'He pretended to be terribly shocked: "Where are you going to?" but I did not tell him. I started to think something here is not kosher', Pim said after the war.[40]

Things took a turn for the worse shortly afterwards. Menachem Pinkhof, who took some clothes for Joop Westerweel to Den Bosch in connection with the escape plans, was arrested at the train station by six German agents on 22 April. Apparently, there were so many agents because they thought that he was armed. Almost at the same time, there was a raid on Tine Segboer's house, where a meeting was going on. Tine, Mirjam Waterman, 'Amo' Rosenbaum, 'Karel' and others, including Henk Brusse, were arrested.

The betrayal could have had even more serious consequences. Jan Smit and Lore Durlacher, who should also have been present at Tine's house, had gone to Westerbork to facilitate an escape and therefore escaped by the skin of their

---

39 GFHA, Cat. no. 143, Interview Yigal (Kurt) Benjamin-Pim van Maanen, 21 september 1989, 1–7. According to Pim van Maanen, Karel told him that he had escaped from Camp Vught. Other sources make no mention of this. See also GFHA, Cat. no. 46, Heinz Durlacher, Manuscript: Stationen des Weges, probably 1945, 1–68.

40 GFHA, Cat. no. 252, Verklaring Mirjam Pinkhof-Waterman, 1957, 1–6. See also Draft text of interview Mirjam Pinkhof-Ellen Landweber, 14, 1988.

teeth. Pim van Maanen also was gone in the nick of time; he had appeared at the meeting briefly to say that he had to leave immediately, because his pregnant wife Gerda was ill.[41]

Fig. 25: Menachem Pinkhof around 1940

The Consequences of Betrayal

Menachem Pinkhof was interrogated immediately after his arrest for seventeen hours. He was handcuffed all this time, but was not mistreated and was also permitted to eat and drink twice. He stated later that he had prepared a 'tactical' story, whereby he told as little as possible that the SD did not already know from 'Karel' or others about the Westerweel group. He admitted that he had

---

[41] Interview Sytske De Jong-G. Loeff, 26 oktober 1998, 14–16. See also GFHA, Cat. no. 225, Verklaring Jan Smit, ca. 1957, 2, 3.

taken Palestine pioneers across the Belgian border twice and had different types of false papers.

After the interrogation, the Germans told him that he would be executed, but this did not happen. Instead, he was sent for ten days to the prison in Haaren in North Brabant. Here, he was locked up, but not interrogated. He was taken to Westerbork at the beginning of May 1944, together with Mirjam Waterman, via the *Huis van Bewaring* in Den Bosch. She was interrogated there by the *SD* man Schönfeld, who treated her well. She admitted only that she had been in hiding and was Jewish.

Menachem Pinkhof, who got through the interrogation reasonably well, reported in his letters from the camp that this result depended greatly on the behavior of person being interrogated. 'You have to be calm, not excited; don't think that you are heroes.' In the same level-headed tone, he also gave several bits of sensible advice to the members of the Westerweel group who were still free. Besides addressing all sorts of practical matters he emphasized that they should not take any risks. They had to do their normal work, i.e. care for people in hiding. 'No more extra efforts, no stunts, no heroic things'.

According to Menachem, it was precisely these extra activities that had led to the affair with 'Karel', who had been very sophisticated in his approach, and they should have mistrusted him much more. 'If we had been calmer, cooler and had had less of a frenetic attitude bent on helping immediately, this would not have happened.' It was mainly a sort of exhaustion which affected all illegal workers that had caused their reactions, he reasoned.[42]

Menachem and Mirjam were placed in the punishment barracks in Westerbork. Because their resistance activities were well known, all sorts of relations did their best to get them out—in any case to avoid their immediate transport to Auschwitz. This was successful. Kurth Blüth, the prominent member of the Jewish Council, discussed their case personally with commander Gemmeker and was able to free them from the punishment barracks. From Amsterdam, people like Carl Oroszlan were able to put them on the Red Cross list with Palestine certificates. Mirjam and Menachem were thus able to go to Bergen-Belsen on 19 May 1944. They were even allowed to take someone with them from the punishment barracks: Heinz Fränkel, the former *Transportführer* of the *Organisation Todt*. Their 'salvation', which was probably due to bribes given to certain high

---

42 GFHA, Cat. no. 173, Brief Menachem Pinkhof smuggled out of Westerbork at the beginning of May 1944.

placed Germans or organizations by Mirjam's wealthy father (who was in hiding), was regarded as 'a unique case' in Westerbork.[43]

In Bergen-Belsen, they were similarly lucky. They knew many prominent people there who helped them get jobs that were not too bad. Mirjam was given work in the old people's and the orphans' home, Menachem went to the so-called 'shoes unit'. They were both in the infamous transport that left for Theresienstadt in mid-April 1945, but was liberated by the Red Army in Tröbitz. They did not stay there for long. Abel Herzberg and several others, who were also in the train, wrote a note about the very poor health situation of the deported Jews in Tröbitz, which their Russian liberators had done little to improve. Menachem and Mirjam took the letter to American troops by bicycle, and they acted quickly. In June, Mirjam and Menachem were back in the Netherlands, where they immediately started rebuilding the *Hechalutz* movement. They left for Palestine in April 1946.[44]

In contrast to Menachem Pinkhof, 'Amo' Rosenbaum, who was also arrested in The Hague, was mistreated during his interrogation. He was then taken to Westerbork as a 'punishment case', just as Ruth Direktor and Thea Perlmutter, who had been arrested at Budel together with Joop and Bouke. All three were deported to Auschwitz a short time later. Ruth and Thea remained in Auschwitz for only a few days and then started on a horrendous journey along factories and work camps in Poland, Germany and Czechoslovakia. In April 1945, the Red Cross took them, exhausted and sick, to Sweden.[45] 'Amo' Rosenbaum also survived the war, but was heavily wounded in the liberation, probably by an American bombardment.

Pim van Maanen and his pregnant wife Gerda went into hiding after the arrests at Tine Segboer's home. They escaped by chance during a German raid. After this, they stayed in Soest. Here, they hid with a Christian woman who came from the Dutch East Indies. This woman had served a prison sentence of a half year in Camp Vught shortly before because of offering help to Jews. With the words: 'the dear Lord has placed you on my doorstep', she hid the Van Maanens and later another eight Jews in her house. Gerda gave birth to the baby in Soest, but the child died a day later from kidney toxicity.

Because they thought that it was too busy and too dangerous in Soest, Pim and Gerda went to Amsterdam. Here, they found a place to hide with Lore Durlacher and Jan Smit. Pim and Gerda continued their forgery work as much as

---

43 Richard Stern, "Joop Westerweel," in *Levend Joods Geloof*, mei 1984.
44 GFHA, Cat. no. 173, Verklaringen Pinkhof.
45 GFHA, Cat. no. 171, Brief Thea Perlmutter aan Kathy Muller, 12 mei 1945.

possible at all the addresses where they hid. After the liberation, Gerda had to be admitted to a psychiatric clinic for some time.[46]

The Germans took Bouke Koning from Eindhoven police headquarters (where he had been taken after his arrest in Budel and where he had once briefly met Joop Westerweel) to Camp Amersfoort for further interrogation. After more than two months, he was transported to Camp Vught, where the circumstances were slightly better. There, Bouke heard that Joop was imprisoned in the Bunker; he regarded this as 'a very bad sign'.

In Vught, he met the Rotterdam resistance man Henk Brusse, who had been arrested during or shortly after the German raid on the Segboer house in The Hague. Bouke knew his name as a helper of the Westerweel group, but had never met him. They decided to be together as much as possible and to support each other. At the beginning of September 1944, shortly before the liberation of Camp Vught, the two of them went with a large transport to the Sachsenhausen concentration camp near Berlin. The circumstances there were very bad. They were then transported to the Gross-Rosen concentration camp in Silesia, where they had to work in a quarry. Here, Henk Brusse, a slender man who was not used to heavy labor, died from exhaustion at the end of December 1944. With the advance of the Red Army, Bouke was transported in an open goods wagon to the infamous Camp Dora in the Harz Mountains, where the Germans manufactured V1 and V2 rockets. The advance of the American army led to many prisoners being compelled to walk to Ravensbrück. In Malchow, a sub-camp of Ravensbrück, a totally weakened Bouke Koning was liberated by the Russians at the beginning of May 1945.

Via Brussels and Breda, he ended up at the University Hospital in Utrecht, where he had to stay for more than a year because of tuberculosis. Bouke was not declared cured until ten years later.[47] During this time, he had received his teachers training diploma and became head of a primary school. Some years later, he had to take early retirement because of a war trauma.

The Final Days of Joop Westerweel

After his failed escape attempt, Joop Westerweel was discharged from the hospital barracks and locked up in the men's section of Camp Vught. Here, he had eye

---

46 Interview Sytske De Jong-G. Loeff, 26 oktober 1998, 20–26.
47 GFHA, Cat. no. 117, Interview Mirjam Pinkhof-Bouke and Froukje Koning, 20 and 30 april 1990, 4–6 and 1–4.

contact with Wil. The Germans had put her to work in a gas mask factory in Den Bosch. However, when she heard that her husband was in Vught, she pretended to be sick, to be able to return to the camp. 'He stood in the corridor of the men's barracks in front of one of the closed windows, I was in the women's barracks. There was a courtyard of about 25 meters between us. We could only look at each other and hold our hands up.'[48]

Shortly after this, the Germans locked Joop Westerweel up in the Bunker, the cell complex of the camp that was housed in an old barracks. Because he was interrogated again, this internment was probably connected to the arrests in Paris, at the end of April 1944 of Kurt Reilinger, Willy Hirsch and others, which are described in the next chapter.

The Bunker cells were cold and damp, brightly lit at night, and dark and somber during the day. The whole situation was set up to break the prisoner's will. For Joop, these were circumstances that led him to reassume his old role of inspirator and organizer. He fashioned chess pieces and checkers out of bread, organizing competitions, whereby the moves were announced through the open windows. He also lectured about politics and literature and read poems written by himself and others.

The poem: 'Avond in de cel' (Evening in the cell), that he wrote in July 1944, appears to express his mood of resignation, acceptance of his fate, yes even luck. The blind wall, lit by the setting sun, the symbol of the German execution place, determined 'the order around our lonely place'. But it could no longer harm Joop. Once again, he saw all the friends who had stood by his side and with whom he had travelled 'the right path'. That was the only thing that counted.[49]

There were many resistance fighters in the Bunker who were sentenced to death and were waiting their sentence to be carried out. Prisoners were regularly executed on the shooting range a few kilometers outside the camp. After the failed attack on Hitler of 20 July 1944, the *Niedermachungsbefehl* was issued. This meant that resistance fighters no longer had to be convicted first, but could be immediately killed after their arrest.

In the Netherlands, the SS officer Dr. Erich Deppner, the head of the *SiPo/SD* Department IV (tasked with fighting the resistance) was charged with formulating lists of prisoners to be executed. For this purpose, he went to Vught, where 1,500 prisoners had been taken from the Scheveningen prison after the Allied invasion in Normandy at the beginning of June 1944. Deppner selected his victims

---

[48] Westerweel, Lijn of cirkel, 80.
[49] Willie Westerweel, "Joop," Westerweel, *Verzet*, 11, 12. See also Collectie Ph. Rümke, Herinneringen W.C. De Jong-Weber, 113–122.

from these prisoners and from the prisoners who were already in Vught. The executions mostly took place as reprisals for acts of resistance elsewhere.

On 11 August 1944, it was Joop Westerweel's turn. A group of printers locked up in the Bunker, who had to make the posters with the names of the prisoners to be executed, let it be known the evening before that Joop and twelve others would be executed the next day. Wil, who was somewhere else in the camp, also heard of the fate that was awaiting Joop. Dr. Steyns, who was imprisoned in the Bunker as punishment for his help to prisoners, suddenly heard 'the beautiful voice of fellow prisoner Co Homme sounding through the building with a song that told us that one or more of our comrades would be killed again tomorrow. Joop knew this sign just as everyone else. He therefore knew what was awaiting him the next morning at 4 o'clock. And then, I dreamed or was this reality? With a firm voice and without any hesitation, Joop read aloud the beautiful poem by Campert for 'de achttien doden' (the eighteen dead).[50]

The End of the Westerweel Group's Activities

The arrests of March and April 1944 left deeper damage than those of autumn 1943 because almost all active members of the Westerweel group were involved, with the exception of courier Tinus Schabbing. Because 'Karel' had also been to Jan Smit's rooms in Amsterdam, he had to find another address to hide with Lore Durlacher, who lived with him. Frans Gerritsen, who had not been present during the discussions about Joop's escape, but whose name had been mentioned, also had to leave his home in Haarlem. He considered that there was a good chance that the *SiPo/SD* would beat more information out of the detainees.

But this did not happen. Shortly afterwards, the Germans did find information on him during the arrest of one of his contacts in Hilversum—outside his work for the Westerweel group—to whom he had supplied false papers, In May 1944, Frans went first for some time to family in Zeist. He then went into hiding in a garden shed with his wife Henny. He made another hiding place under this shed. Their twins went to live with his parents.[51]

At this time, Norbert Klein had been housed for a while with Bob Jesse and his wife Dientje. He was nursed there by the 24-year-old Lilo Spiegel, from Berlin, who had previously worked in *Het Apeldoornse Bos*. In mid-April, she left for France, accompanied by Max Windmüller. Henny Gerritsen travelled with

---

50 Collectie Rümke, Ibid., 37, 38.
51 GFHA, Cat. no. 63, Brief Frans Gerritsen aan 'Beste allemaal', 10 september 1945, 1, 2.

them to the Belgian border and took their Dutch papers back for reuse. Lilo was not the last pioneer who left for France. At the beginning of May 1944, Ruth and Heinz Durlacher followed. Shortly before, Max had taken their four-year-old son Uri to Paris. Uri had been in hiding since 1942, separated from his parents.[52]

After the war, Jan Smit and Tinus Schabbing stated that the Westerweel group's work had basically ceased after the summer of 1944. It is better to say that the character of the activities changed. As a result of the arrest in Paris of several *chalutzim* leaders, of whom there is more later on in this book, the escape route to France was cut off in the late spring of 1944. The invasion in Normandy at the beginning of June and the rapid Allied advance through France and Belgium, followed by the liberation of the southern part of the Netherlands in the autumn of 1944, changed the situation completely. The German capitulation had now become a question of time.

What one could call the Westerweel group's 'normal' activities, such as supplying ration cards to people in hiding, were continued by Jan Smit, Tinus Schabbing, Harry Asscher and Elly Waterman, Mirjam's sister. Frans Gerritsen continued to forge papers and, as described, was also active around Westerbork. In addition, he built a large number of hiding places for Allied pilots and for others—both alone and together with Tinus Schabbing.[53]

After this, the railway strike that was called by the Dutch government in mid-September 1944 made transport virtually impossible. Caring for pioneers in hiding who were mainly in the provinces of Friesland and North Holland continued, but that work was co-ordinated by Carli Oroszlan. Harry Asscher, Elly Waterman and the local resistance were important helpers.

From the end of 1944 and the beginning of 1945, the food supply in the west and middle of the country caused by the *Hongerwinter* required all their attention. This also applied to the members of the Westerweel group who were still free. (The *Hongerwinter* was a famine in the German-occupied part of the Netherlands, especially in the densely populated western provinces).

During this time, Frans Gerritsen made a grueling journey with a homemade cart, but all the food collected underway was stolen. Later, there were other journeys to Overijssel and Drenthe, with a better result. Jan Smit and Tinus Schabbing brought food and wood for cooking by boat from Utrecht to Amsterdam.

---

[52] GFHA, Cat. no. 230, Verklaring Lilo Spiegel, december 1955, 1–4. See also GFHA, Cat. no. 46, Verklaring Heinz Durlacher, 1956, 1–3.
[53] GFHA, Cat. no. 63, Verklaring Frans Gerritsen, 1–3, ca. 1957, aanvullingen september en oktober 1989.

These goods were divided over different addresses. During the final months of the war, Jan Smit also worked for the group publishing the left-wing socialist newspaper *De Vonk*.[54]

* * *

The Germans were finally able to largely dismantle the Westerweel group in different stages. Of the people who initiated the 'hiding operation' in Loosdrecht August 1942 and had remained active afterwards, only Jan Smit was still free two years later. Moreover, the group shared this fate with a large number of other resistance organizations. The causes of the arrests were a combination of German detective work, betrayal, coincidence and carelessness. However, when the group was forced to end or adapt its activities by circumstances, it had largely completely its mission: bringing as many Palestine pioneers as possible to safety.

With regard to the first two aspects of the German action, detective work and betrayal, which one might call the Siamese twins of fighting the resistance, the tactic of infiltration was an almost unknown phenomenon in the Netherlands. The resistance had to get used to people such as the 'married couple' De Ruiter, who portrayed themselves as helpers only to betray the Jews they were hiding. Also, the brutal interrogation methods used by the Germans, whereby people were mistreated for as long as it took until they provided the information demanded, were unknown in the Netherlands. Joop Westerweel had to experience this physically.

With regard to the carelessness mentioned, one could say that the Westerweel group never really started to act more professionally during the roughly two years that it existed.

Leader Joop Westerweel often appeared to be overly trusting and was sometimes so overconfident (for example, the dinner in Rotterdam or the discussions with German army soldiers in the train) that it is frankly amazing that he was not arrested earlier. Yet, even his more cautious comrades-in-arms Jan Smit, Menachem Pinkhof and Mirjam Waterman failed to perceive the infiltration of 'Karel'.

You could call Joop Westerweel's position during his imprisonment tragic. From the beginning of the war, he had been convinced that he would have to endure a heavy time, whereby he would stay at his post 'even in the toughest times'. But his arrest and the ensuing mistreatment left him totally dejected. Con-

---

54 GFHA, Cat. no. 63, Brief Frans Gerritsen aan 'Beste allemaal', 10 september 1945, 10–15. See also GFHA, Cat. no. 225, Verklaring Jan Smit, ca. 1957, 2, 3. Interview Sytske De Jong-Tinus Schabbing, 13 oktober 1998, 14.

tinually searching for possibilities to escape, whereby he also involved members who were still free, was the main reason for the disappearance of the group, of which he had been the inspiration and driving force for almost two years. Only after the arrests in Den Bosch and The Hague and his imprisonment in Vught did Joop Westerweel appear to accept his fate, which he finally underwent very courageously.

11 *Les Hollandais*, the Westerweel Group in France

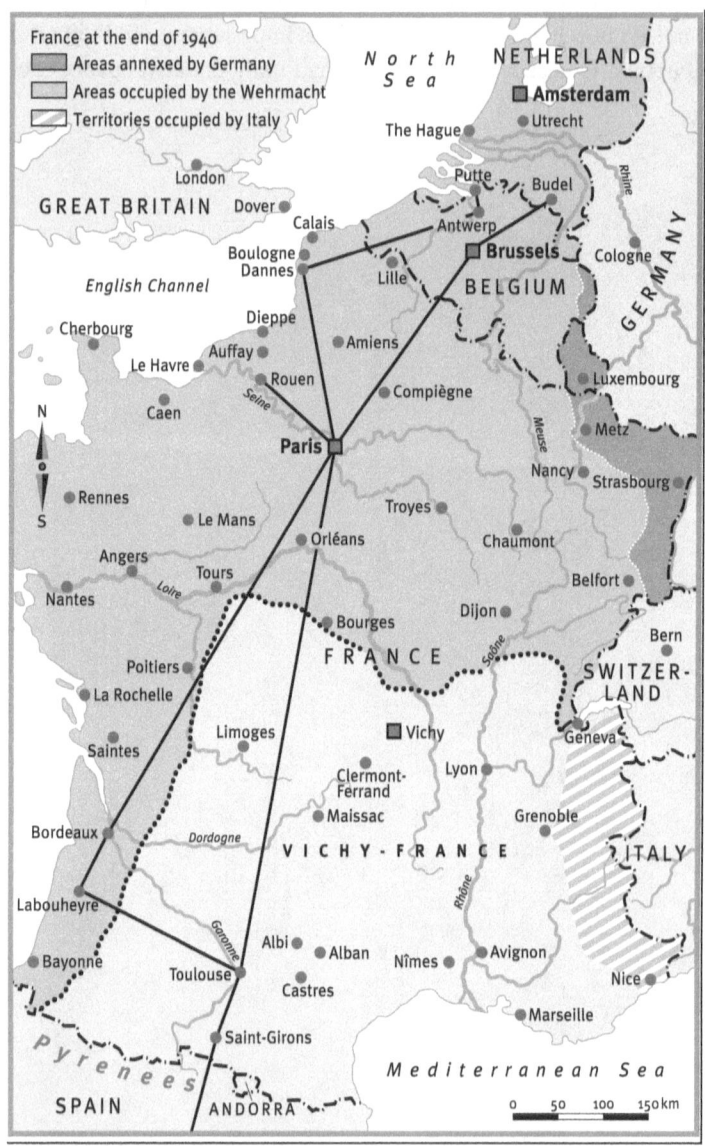

Map 2: The Palestine pioneers in France: escape routes, maquis, OT worksites

Earlier in this book we described how the Palestine pioneers found a difficult route to the French coast to work for the *Organisation Todt* on the *Atlantic Wall* and other projects. During 1943 and the beginning of 1944, there were about 150 pioneers in France, most of whom intended to continue on to Spain.

Cooperation with the Jewish Resistance

On his departure at the beginning of August 1943, Kurt Reilinger (Shushu Simon's successor as representative of the pioneers in France) was given two tasks by the board of the umbrella organization *Hechalutz*. He had to contact the pioneers who had left earlier and also restore Simon's former relations with the Jewish resistance in France.

For this first assignment, Kurt visited a large number of pioneers working in France in August and September 1943. In addition, he established a sort of 'headquarters' for the Dutch *chalutzim* in France at varying hotel addresses in Paris. With the cooperation of people like Willy Hirsch, Alfred 'Zippi' Fränkel, Hans Ehrlich and Max Windmüller, the newly-arrived pioneers were offered sanctuary and given further guidance. Kurt and his employees also managed the money for their activities, which came largely from the Netherlands, and provided false travel passes and other documents.

The 'headquarters' was later reinforced by several female pioneers, including Lolly Eckhardt, Metta Lande, Betty Britz (who had also worked in Loosdrecht) and Paula Kaufmann. Lolly Eckhardt's room in the hotel *Versigny* in the Rue Letort then functioned as a base.

Lolly Eckhardt was a 21-year-old Viennese who went to the Netherlands at the end of 1938. She had followed the usual pioneers' path (*Werkdorp*, Amsterdam, *Het Apeldoornse Bos* and again Amsterdam until June 1943). She had been fortunate to escape capture during the large raids in the city center and had gone into hiding in Sintjohannesga, Friesland, with a Catholic blacksmith's family. Harry Asscher and Lore Durlacher regularly visited her there.

At the end of 1943, when Zippi Fränkel told her that there was a possibility of going to France, Lolly agreed immediately, 'because I could no longer be alone'.[1] Zippi had taken her to Frans Gerritsen's house in Haarlem and there she was impressed by the effort and achievements of the people in the Westerweel group.

---

1 GFHA, Cat. no. 47, Verklaring Lolly Eckhardt, april 1956, 1–4.

Lolly, who had almost reported for Westerbork several times, had then resolved 'to also help, in case it was necessary'. In Paris, she worked with other helpers to provide housing and guidance to newly arrived pioneers.

The *Armée Juive*

During October, Kurt Reilinger also moved forward with the second part of his assignment, namely to contact the Jewish resistance in France. The connection was realized through Switzerland, where Shushu Simon's widow Adina was in contact with the prominent socialist and Zionist Marc Jarblum. Shushu had had several discussions with Jarblum in Lyon in the autumn of 1942. Jarblum had fled to Switzerland in March 1943. Through Adina, he referred Reilinger to Avraham Polonski, who was the co-leader of the *Armée Juive* (*AJ*) in Toulouse. This was a regional resistance organization with branches in cities such as Toulouse, Nice, Lyon and Grenoble. On paper, it had a membership of almost 2000, 300 of which were the core members. However, the *AJ* was not a real army; the organization had hardly any weapons or military knowledge.[2]

Polonski was a technician from White-Russia, who lived in Toulouse after a stay in Palestine of more than 15 years. He was a supporter of the right-wing, 'Revisionist' movement in Zionism that was led by Ze'ev Jabotinsky. From the autumn of 1942, the pragmatic Polonski worked in the *AJ* with Aron Lublin, a socialist Zionist, and also a technician, who came from Poland and was active in Lyon.

Polonski was an excellent organizer, but also a lover of romantic rituals. For example, all active members of the *AJ* had to take an oath of allegiance to the leaders of the *AJ* in a dark room that was lit by a strong flashlight to prevent recognition of the other people present. Other attributes used during such rituals were the blue-white Zionist flag and a Torah scroll. At the end of the oath, the prospective members had to say: 'To the revival of the Jewish people and the country of Israel! Freedom or death!'

Lublin was not present at this sort of ritual, but this did not prevent his cooperation with Polonski. The *AJ* helped Jewish refugees, but also had a security service that collected information about traitors of Jews. Later, especially the group in Nice carried out attacks on German targets and collaborators.

In addition, the *AJ* was active in searching for a route to Spain. The idea for this could have come from the proposal that Shushu Simon had made in the autumn of

---

[2] Anny Latour, *La Résistance juive en France (1940–1944)* (Paris: Stock, 1970), 105, 106. See also Lucien Lazare, *La résistance juive en France* (Paris: Stock, 1987), 115–117.

1942. This plan, which made it possible for young Jews to go through Spain and enter military service with the Allies to fight the Nazis, appealed to Polonski. In his view, it would contribute to the idea of a Jewish rebirth in Palestine.[3]

Fig. 26: Members of the Armée Juive, with Avraham Polonski in the centre, in a post-war meeting with the Israeli Prime Minister David Ben-Gurion, to his right

Jacques Roitman and the Road to Spain

In December 1942, the *AJ* organized a crossing to Spain. Twelve members of the Jewish scout movement crossed over the Pyrenees. Two of them were, however, arrested by the Germans and deported to Poland. Partly because of the poor safety and the harsh circumstances, the *AJ* decided not to undertake any further journeys. At the beginning of July 1943, the organization ordered Jacques Roitman, a youthful, religious Jew from Poland, to find an escape route over the Pyrenees. At the time, he did not have much more information than the address of the mountain guide 'Adrien'. This guide had said that he was prepared to take refugees to

---

3 Lazare, *La résistance*, 306, 307.

Spain for a great deal of money: between 5,000 and 10,000 francs per person, depending on the size of the group.

Roitman then busied himself with the financial and logistical organization. The *Joint*, the influential and wealthy American-Jewish aid organization, bore the lion's share of the costs via Switzerland. Their money was used to pay the guide and provide other necessities, such as food and warm clothing. With regard to the logistics and safety, it was important that people aiming to go to Spain stayed only a short time in Toulouse, where the *AJ* headquarters were. In October 1943, everything was ready for the first crossing, with which also Dutch pioneers, *les Hollandais* (as they were called by the *AJ*) would go.[4]

Kurt Reilinger travelled with this news to Auffay in Northern France in mid-October 1943. Hans Flörsheim, who then worked there, provided a report of the meeting in his book. After briefly describing the position of the Jews in France, Kurt began to describe the possibility of an escape route to Spain with the help of a Jewish sister organization, as he called it. The undertaking would be no easy feat, he warned. Flörsheim's enthusiastic reaction: 'That was a nice surprise for us! In Spain we would be free of these false names and papers!' However, Flörsheim soon realized that this enthusiasm was not shared by everyone. Five of the roughly twelve-man group in Auffay applied immediately for the journey over the Pyrenees. The others apparently wanted to see how things were organized.[5]

The hesitation to sign up for a journey to Spain was understandable. As described, the circumstances in France were much more favorable than in the Netherlands. Moreover, it hardly seemed to interest the Germans that there were Jews among the *Organisation Todt* workers. So why take great risks to go to Spain?

There were, however, also arguments for leaving; the Jews were threatened in France as well. Tens of thousands of mainly foreign Jews were deported to Poland from the French 'Westerbork' Drancy in 1942 and 1943. The Germans were unpredictable. The indifference as to the Jewish background of the *OT* workers could easily turn into persecution. There were also rumors that, in the event of an Allied invasion, the Dutch workers would have to go to Germany, with all the associated dangers.[6]

The Zionist ideology also played a part in the background to this discussion. In 1943/44, it was far from certain that the British would keep their promise in

---

[4] Centre de documentation juive contemporaine (CDJC), Paris, DLXI (4), Archief Jac. Roitman, 1–10, 16 oktober 1974. Other resistance groups also organised crossings to Spain. Kurt Benjamin and Kurt Ehrlich reached Spain in September 1943.
[5] Flörsheim, *Über die Pyrenäen*, 86.
[6] GFHA, Cat. no. 46, Heinz Durlacher, Manuscript: Stationen des Weges, 54–56.

the Balfour Declaration of 1917 to support the establishment of a 'National Home for the Jewish people' in Palestine, after the end of the war. Every Jew who lived there could then contribute to the realization of that Home.

Shortly after the meeting in Auffay, there was a discussion in Saintes (a small place between Bordeaux and La Rochelle) between Kurt Reilinger and several representatives of the pioneers who worked on *OT* projects on the south-western coast. The people present included Emil Glücker, Erich Sanders and Berrie Asscher. Kurt also described the possibility of going to Spain, but now mentioned the condition that every candidate had to promise that he would join the 'Jewish army' upon arrival in Palestine. This condition was apparently made by the *AJ*.

It was unclear what was meant by the 'Jewish army'. It could have indicated the *Haganah* (the Jewish self-defense organization, founded in 1920) or the *Irgun* (a right-wing nationalistic splinter group), which was linked to Jabotinsky's Revisionists. None of the people present wanted to join the *Irgun*. Kurt Reilinger would investigate this further.[7]

An Initial Start

In Saintes, Kurt also told them that the first journey to Spain would soon take place. Berrie Asscher and Heinz Meierstein would go with this first crossing. If everything went well, the others would follow. Berrie travelled to Toulouse, the departure point, on 3 November 1943, provided with a false German travel document. After some delay, it was confirmed that the journey would start a week later.

The day before, the ten pioneers (who had in the meantime gathered in Toulouse) were required to take an oath. Possibly through the interference of Kurt Reilinger, this was not a promise to serve in a Jewish army, but an oath of allegiance to the *AJ*. Berrie Asscher thought that the ceremony in the dark room with the flag and the Torah resembled a scene from a film about the *Ku Klux Klan*. He and the others regarded it as a kind of sacrifice that they had to make to ensure cooperation with the *AJ*.[8]

On 10 November 1943, the Dutch pioneers travelled by train and bus from Toulouse to a little village near Saint-Girons in the department of Ariège at the

---

[7] Asscher, *Van Mokum*, 160, 161. A Jewish Brigade in the British army was not founded until autumn 1944.
[8] Asscher, *Van Mokum*, 164. Some pioneers refused later to take the oath. However, they were allowed to go anyway.

foot of the Pyrenees. There, they met the guide Adrien, with whom they continued on to a primitive shepherd's hut. Here, they had a five-day wait for other travelers. These included several pioneers and also some older participants, probably also Jews, from countries such as Czechoslovakia.

The group of 21 men started the climb after sunset, led by Adrien, who was armed with a pistol. This quickly became too much for one of the older participants and he returned to the hut. After walking for several hours, the participants reached a second shepherd's hut. However, they awoke the next morning to see that a half a meter of snow had fallen that night. Adrien carried out a short reconnaissance, determining that the route was impassable. He also said that he had seen bear tracks. The group had to return to Toulouse. Adrien stated that a new attempt could not be made to cross the Pyrenees until three months later.

The people trying to reach Spain were now faced with the problem of how to pass this time in a safe and meaningful way. An *AJ* representative suggested joining the *Maquis*. This was a collective name for small mobile resistance groups that carried out attacks on German troops and transports in France from thinly-populated, mountainous areas. However, most pioneers, who often spoke hardly any French, did not like this idea very much. Hans Flörsheim, who felt unsafe in France, decided together with a comrade from Gouda to try to flee to Spain by hiding under a train. They travelled to Bayonne and made several attempts, but realized that this escape route was not possible because the Germans inspected also the undersides of the trains.[9]

After they returned to Bordeaux in December 1943, they decided to again work for an *OT* project. Shortly afterwards, they met Christiaan Lindemans, a Dutch illegal worker, who smuggled Allied pilots to Spain. He promised to also take them across the border. However, after several meetings, including one in Paris, Lindemans, who would later become known as the infamous double agent King Kong, remained unreachable.

Despite warnings from Kurt Reilinger and Willy Hirsch to be careful, Hans Flörsheim continued trying to contact Lindemans. This led to Flörsheim's arrest in mid-February 1944 and internment in the prison of Bayonne. He was not released until six weeks later.[10]

---

9 Flörsheim, *Über die Pyrenäen*, 100–105.
10 Flörsheim, *ibid.*, 109–115, 126–156. Hans Flörsheim was not betrayed by Lindemans. Apparently, however, the Germans had then already infiltrated Lindemans' network.

Adventures in the Maquis and Italy

After the roughly twenty remaining pioneers of the failed attempt to get to Spain had stayed in Toulouse for a few weeks, Kurt Reilinger and Willy Hirsch thought that this attracted too much attention and was dangerous. They again asked for volunteers to join the *Maquis*, after which six men reported for duty, including Berrie Asscher.

During 1942, various Jewish groups in France started to join in the armed fight against the Germans. The *Main-d'Oeuvre Immigrée* (MOI, migrant laborer) groups, which consisted largely of Eastern European Jews that were linked with the communist party, played the most important part here. Other groups were active in general resistance organizations.

In autumn 1943, the *AJ* started to work with a *Maquis* unit near Albi in the mountainous department of Tarn. The intention was that the Jewish resistance fighters would be given military training, after which they could form their own units or join the Allied troops in Palestine via Spain. The Dutch pioneers formed the first unit that followed this practical training. At the beginning of December 1943, the six of them travelled from Toulouse by train and bus—and after a hard march reached a deserted farm in the tiny hamlet of Alban, where they were quartered. Because of suspected anti-Semitism among some of the French resistance, they were not allowed to say that they were Jews. The thinly-populated and remote, wild region was '*Maquis* country', where the Germans appeared sporadically—and then only in armed convoys. The fledgling *Maquisards* were given a rifle, a uniform and military training, which consisted of weapons instruction and standing guard exercises. There was no shooting practice because of the limited stock of ammunition.

Because of the threat of German attacks, the unit (which had about twenty men and was led by a French army lieutenant) had to move several times. Finally, a completely Jewish group was formed in a deserted farm with the name of Le Bic or Biques. This group was led by Meijer Sal, a rabbi who had been a non-commissioned officer in the French army. The Dutch pioneers group, which now had ten men, stayed there until 7 February 1944, when word reached them that a crossing to Spain was again possible. They left for Toulouse, which put an end to the *Maquis* adventure.

Together with other guerrilla groups, the Jewish *Maquis* fighters would later play a significant part in the liberation of South France. They were involved in operations against German troops in the small town of Castres and in actions in cities such as Toulouse, Limoges and Lyon.[11]

Fig. 27: A maquis unit of the Armée Juive in the department of Tarn in 1944

The cooperation with the *AJ* did not remain limited to the *Maquis*. One of Reilinger's assistants, Alfred 'Zippi' Fränkel (who for some time had been a courier for the Jewish Council to Westerbork in the Netherlands) was asked to go from Toulouse to the Franco-Italian border area in the spring of 1944. Several groups of Jewish refugees were staying in this area, which was controlled by Italian partisans. Their position was, however, threatened by German troops and Italian fascists. Zippi's main task was to restore contact with the refugees and, in his own words, 'to assess the state of children and older people, and help them.' He took money with him to give to the families in need.

Zippi Fränkel travelled to Italy and was able to find the groups of refugees. He made a deal with the partisans that they would help the refugees in exchange for financial support. He urged the young men among the refugees to either go to

---

[11] Lazare, *La résistance*, 125–129, 312–327. See also Latour, *La Résistance*, 211–214. Asscher, *Van Mokum*, 172–179.

Spain or join the *Maquis* in France. He reported his findings to the *AJ* in Toulouse.

During his second visit, he became involved in fights between partisans and German soldiers. He warned the refugees and partisans of German actions and then travelled through to Florence and Rome to pick up information about other Jewish refugees. On his way back, Zippi, who had falsified German police papers, was arrested by Italian fascists, who accused him of espionage. He called in the help of German soldiers and succeeded in getting free. Zippi then travelled back to Toulouse and Paris in mid-April 1944 to report on his activities.[12]

New Chances for Spain

The *Maquisards* who returned to Toulouse at the beginning of February 1944 were initially confronted with new delays. In order not to be conspicuous in the city, they therefore had to spend another two weeks to Paris. At the end of the month, however, the time was right and a large group of 34 people, including 23 *chalutzim*, met up near the Pyrenean village of Saint-Girons. Besides the pioneers, several French police officers, two Allied pilots and several older French Jews were also in the group. The guide was again Adrien.

The group, which as described earlier, was spoken to by Joop Westerweel before leaving, had food for two days. According to Adrien, that should have been sufficient. However, the weather was the biggest challenge, the main problem being the biting cold temperature of about minus 15 degrees celcius. After a horrendously heavy journey of not two, but five days (during which the older French people turned back and several participants suffered frozen limbs), the group finally reached their destination on 4 March 1944. Group member Paul Siegel wrote later that he had the idea that he was reborn upon his arrival in Spain, more than a month after his escape from Westerbork.[13]

Anyone who thought that the journeys over the Pyrenees might go more easily after this first successful crossing, given the better weather conditions in the spring, was wrong. Apart from the weather, the behavior of the guides was a problem. Adrien, for example, made errors of judgment—due in part, perhaps, to his heavy drinking—and did not keep agreements. This sometimes led to great danger for the participants in a crossing. For example, in violation of his promise, he left Paul Siegel and Berrie Asscher's group before the border with

---

12 GFHA, Cat. no. 55, Verklaring Alfred Fränkel, 6–9.
13 Siegel, *Locomotieven*, 164.

Spain without clearly explaining the rest of the journey. It was only as a result of swift action taken by a French policeman with mountain experience that there were no accidents.

Adrien separated from the group again during a later journey. This group ultimately got lost and Israel Tiefenbrunner, a Loosdrecht pioneer, was fatally injured. Adrien himself was killed in a shoot-out that occurred during the preparation of a new crossing in the spring of 1944—by either Germans or collaborators with the *Milice* (the French militia).[14]

Other paid guides were also sometimes less reliable. End February 1944, an attempt to cross near Pau failed because a second guide, who was supposed to take the group over, did not turn up. This might have been because the group, to which Lore Sieskind and Hans Flörsheim belonged, had been delayed. The situation did not improve until April 1944, when Spanish Republican exiles, opponents of Franco, started to act as guides with the transports. They took several groups, which also included pioneers, across the border at Andorra, without payment.

Uri Durlacher had a remarkable crossing. Uri was the three-year-old son of Heinz and Ruth Durlacher. His parents fled to the Netherlands as husband and wife in mid-1939. They had followed a *hachshara* course in Germany, and wanted to continue it in the *Werkdorp Wieringermeer*. Uri was born in November 1940. After the *Werkdorp* was closed, Heinz worked near Deventer for some time. When the persecution of the Jews started in May 1942, he and Ruth tried to find an address for Uri to hide. However, they did not succeed until five months later, when they found an address in The Hague.

They decided to also go into hiding in May 1943, but Heinz Durlacher was arrested at the Amsterdam Central Station and locked up. Shortly afterwards, however, he was able to escape from a transport to Westerbork. Ruth and later also Heinz then went into hiding with a brother of Ans Roos in Utrecht. In April 1944, Kurt Reilinger suggested taking Uri to Spain. After much deliberation, his parents agreed. Max Windmüller picked Uri up in The Hague and took him to Belgium via the 'green border' and on to France. In Toulouse, he was housed in a children's home, where there were more Jewish children. At the end of May 1944, Uri took part in a children's transport, whereby older children carried him in turn on their backs to Spain. Two months later, his parents arrived in Spain and they were reunited.[15]

---

14 Siegel, *ibid.*, 214, 215.
15 GFHA, Cat. no. 46, Verklaring Heinz and Ruth Durlacher, april 1956, 1–3, 1, 2.

There is no complete overview of the number of Palestine pioneers that succeeded in crossing the Pyrenees to reach Spain. In his book about the Jewish resistance, Lucien Lazare mentions a number of more than 90. However, this appears too high. Dutch sources assume a number of roughly 60 pioneers, with the group of 23 men to which Paul Siegel and Berrie Asscher belonged as its core. Later, smaller groups or individual pioneers arrived.

The number of 60 probably does not include Moshe Osterer's group of four men plus Kurt Benjamin and Kurt Ehrlich, who arrived in Spain in the spring and in mid-1943, respectively. Several other pioneers may have possibly also reached Spain on their own. With them included, the total number of people trying to reach Spain could have been around 70.[16] This represents about half of the number of pioneers that left the Netherlands to go to France in 1943 and 1944.

Shelter for Pioneers in Spain

The previously-mentioned group of 23 Palestine pioneers, who left from Saint-Girons at the end of February 1944, was arrested by the Spanish police upon arrival at the beginning of March. The fate of this group, about which there is more information, is described here because their story is largely representative of the experience of other *chalutzim* in Spain. The Dutch consul in Barcelona provided the initial assistance, after which they were given permission to contact Dr. Samuel Sequerra. He was the Spanish representative of the *Joint*, the American-Jewish aid organization. He ensured that the group with Berrie Asscher and Siegel could stay in hotels in Lerida and would receive money for clothing and daily expenses.

Sequerra probably also arranged to prevent all male pioneers between the ages of 18 and 40 years from being locked up in one of the infamous internment camps, such as Miranda del Ebro. Many pioneers feared such a destination. After consultation with the police, only a few pioneers needed to report.

Besides Sequerra, Fritz Lichtenstein, a representative of the *Jewish Agency* that organized the immigration to Palestine, was also active in Spain. During a meeting, he told the group that the Agency was busy organizing a boat journey to Palestine. As soon as sufficient potential passengers arrived, a ship could be chartered to take them. The Palestine certificates required by the British govern-

---

[16] Lazare, *La résistance*, 311. See also Van der Straaten, *Om nooit te vergeten*, 133. GFHA, Cat. no. 173, Interview Menachem Pinkhof-Haim Avni, 9.

ment would not be a problem. Of the 75,000 certificates that had become available in 1939, only a small number had been used because of the war.[17]

After two months, the Asscher-Siegel group was able to move to Barcelona, where the they were compelled to wait in assigned rooms in cheap guesthouses. Several group members made themselves useful by helping to take care of several dozen Jewish children, some of whom were orphans. These children had arrived in the spring and summer from France along a much easier route, close to the Mediterranean coast. They were housed by the *Joint* in its own children's home.

During July and August 1944, most *chalutzim* obtained the requisite Palestine certificates through the British consulate in Barcelona. In September, a representative of the *Jewish Agency* told them that a Portuguese ship had been chartered for the journey to Palestine. The ship, the *Guine*, would be ready to sail in October.

And so, it happened. At the beginning of October, the Asscher-Siegel group, the children and other people from various places who had travelled to Spain boarded a special train from Barcelona to Madrid, where passengers from other places joined them. The train left the next day from Madrid to Cadiz, in southern Spain, from where the *Guine* could sail. The old passenger ship arrived several weeks later from Portugal and left on 25 October 1944 to Tanger, which was under Spanish authority. Here, a hundred Sephardic Jews from Greece boarded the ship. These people had been admitted to Tanger because of their Spanish origin. From Tanger, the *Guine* left for Haifa, where the ship arrived on 5 November 1944 with more than 430 passengers, including 55 Dutch pioneers.[18]

The guidance described here contrasted sharply with the support, or better said the lack of it, given to Moshe Osterer's group of four men who had arrived in Spain in spring 1943. Initially, they were more or less left to their own devices. Two of them went to the Dutch colony of Surinam via Jamaica, where they were imprisoned for months because they were thought to be German spies. After a short stay, they went to the United States to serve in the American army. Osterer himself, after house arrest for a period of time and a stay of several months in an internment camp, was placed in a guesthouse in Madrid at the cost of the

---

17 Asscher, *Van Mokum*, 196–211. For the organisation of the help in Spain and the associated problems, see: Avni, *The Zionist underground*, 572–590.

18 Asscher, *Van Mokum*, 205–217. See also Siegel, *Locomotieven*, 176–185. Siegel and several other pioneers, who were afraid of internment, tried to get to Barcelona on their own. They were, however, quickly arrested. Two of them had to go to the Miranda del Ebro camp for some time.

Joint. At the beginning of 1944, he boarded the Portuguese ship the *Nyassa*, together with Kurt Benjamin and several other pioneers, bound for Palestine.[19]

Arrests in Paris

The crossings to Spain organized by the *AJ* were successful, apart from the inevitable setbacks and delays. The Paris 'headquarters' of Kurt Reilinger (who had helped to make the journeys possible) was, however, plagued from the end of April 1944 by arrests. This was the consequence of the detention of the Dutch pioneer Joop Andriesse at the Swiss border. At the end of March 1944, Max Windmüller had taken Andriesse (who was born in Utrecht in February 1924) together with his wife Fiet Hess (who was also 20 years old) by train from The Hague to Paris. They travelled on forged leave passes; the journey was organized by Menachem Pinkhof.[20]

Joop Andriesse, who worked under a false identification card as a non-Jewish laborer on a farm in Veghel, Brabant, and also assisted in making forgeries, had initially planned to leave for France in January 1944. This transport had, however, failed because of the arrest of a German deserter mentioned before, who travelled with them. Andriesse was the only person to escape, because he was in another train compartment. At the beginning of April, the journey to France was successful. From Paris, Joop Andriesse and his wife were sent on to Toulouse. However, because the maximum number of participants for the next crossing had been reached, they could not go. Fiet then went to work with Lore Sieskind and others on a farm near Toulouse. Joop travelled back to Paris to help Kurt Reilinger.

Andriesse travelled on to Switzerland in mid-April 1944 with a false *Sonderausweiss D*, a special identity document that gave access to the border region. However, Andriesse had little experience with resistance work and was arrested by the Germans in the French border town of Pontarlier in the department of Doubs.[21] A check of whether he had arrived safely in Switzerland was apparently

---

19 GFHA, Cat. no. 166, Verklaring Moshe Osterer, 4. See also Cat. no. 72, Verklaring Willy Gerler, 1–19, 27 april 1994.
20 The two only entered a religious wedding. This was performed in Utrecht by a Jewish lawyer, who was exempted from reporting as a Jew.
21 GFHA, Cat. no. 4, Verklaring Fiet Andriesse, 1–4, december 1955. See also Verklaring Joop Andriesse, 1–3, december 1955 (Translation from Hebrew to Dutch by Ithamar Lehmann). Andriesse was a prisoner in Besançon until autumn 1944, after which he was transferred to Dachau. He survived the war.

not made. The *SiPo* got the address of the Paris 'headquarters' in the Rue Letort, either in his baggage or through interrogation.

On 27 April, the Germans raided Lolly Eckhardt's hotel room, where, as they apparently knew, a meeting was being held. Besides Lolly, Kurt Reilinger, Zippi Fränkel, Willy Hirsch and Susi Hermann were also present. 'There was a knock, we opened the door and four Gestapo men in civilian clothing appeared with revolvers pointing at us.' The room was searched and all documents that were on a table were confiscated. The Germans also found a receipt for a suitcase left at the station of Toulouse and other papers. A revolver and ammunition were, however, not discovered.[22]

The five of them were imprisoned in Fresnes and continually interrogated. They said they were Dutch workers, and that Kurt Reilinger had organized their papers. However, the Germans gradually found out their real identity and activities through interrogation and physical abuse, with Willy Hirsch as their main victim.

After Kurt Reilinger's arrest, Max Windmüller took over his task, together with several employees of the 'headquarters' who had not been arrested (such as Metta Lande, Hans Ehrlich, Lilo Spiegel and Betty Britz). He sent warnings to all pioneers and gave them other places of work—several of them going to work at a German airfield near Lyon. Max and his helpers also maintained the connection with the *Armée Juive* (*AJ*), so that the crossings to Spain could continue.[23]

The Porel Affaire

In the spring of 1944, Henry Pohorylès, the *AJ* head in Nice, told a non-Jewish illegal worker that his group had a serious shortage of weapons. The worker remembered that an acquaintance of his, Olga 'Lydia' Tcherwinsky, had told him shortly before that her boyfriend was an agent of the British *Intelligence Service* (IS). This friend, who called himself Charles Porel, was in reality the German counter-espionage agent Karl Rehbein. He had a meeting with Pohorylès in May 1944, during which he told him that he was a Canadian agent of the IS.[24]

---

22 GFHA, Cat. no. 47, Verklaring Lolly Eckhardt, 4, 5.
23 GFHA, Cat. no. 48, Verklaring Hans Ehrlich, 2, 3. See also Cat. no. 230, Verklaring Lilo Spiegel, 3.
24 Karl Rehbein was an employee of the German counter espionage. Initially, he said that he was an Austrian who had formerly fought in Spain. He tried to infiltrate into resistance organisations in this way. Later, he also pretended to be a so-called Canadian-born officer of the British secret service. Rehbein defected to the British in the autumn of 1944. This time, he said he

Fig. 28: Willy Hirsch, the organizer of the escape routes in Belgium and France, died in Bergen-Belsen, in 1945.

During the discussion, Porel/Rehbein made some major promises about weapon droppings and told Pohorylès that, after the war, the members of the *AJ* could enlist in the British army, with the possibility of going to Palestine.

Pohorylès informed the *AJ* leaders in Toulouse of the meeting and they gave him permission to continue the contacts. But there was no lack of warnings. Jules 'Dika' Jefroykin, a French reserve-officer who was involved with the *AJ*, told another prominent Jewish resistance fighter about the contacts with Porel. This man said to him 'Dika, cease all contact with this man. Drop him immedi-

---

was a representative of a group of German officers who wanted to talk about peace. The British secret service quickly unmasked Rehbein as a German agent and took him to England, where he was probably kept prisoner in Latchmere House in London until 1947. He was then not extradited to France, but was allowed to settle in Germany, which would not extradite him to France, because he was a German citizen. Besides the betrayal of the *Armée Juive* group, Rehbein was partly responsible for the betrayal and ensuing execution of 35 young Paris resistance fighters in mid-August 1944. He was condemned to death in abstentia in 1952. See https//www.discovery.nationalarchives.gov.uk., Karl Ludwig Rehbein file, kv2/3128, accessed 15 September 2014. See also *L'Humanite*, 24 August 2002. Also: CDJC Paris, Statement by Jeanne Latchiver, DLXI-55, 'Rapport concernant l'affaire Lydia Tchervinsky et Charles Porel', 1974.

ately. You don't believe that the Allies are interested in a handful of underfed, untrained Jews.'[25]

However, AJ leaders Polonski and Lublin allowed themselves to be blinded by the promises of Porel and his helpers, including *'Captain Jack'*, the French traitor Count Guy Glèbe d'Eu, who spoke fluent English. For example, Porel offered to take two *Maquis* leaders, Jacques Lazarus and Rabbi René Kapel, to London. They would be picked up by an Allied airplane from an airfield at Tours. The contacts were continued in July 1944 in Paris, where Max Windmüller also became involved.

Kurt Reilinger had always kept the hiding operations and Spanish activities as separate as possible from the active resistance, such as that of the *AJ*. The limited participation with the *Maquis* and Zippi Fränkel's Italian mission were exceptions.[26] The invasion in Normandy and the ensuing Allied advance had, however, led to a strong increase in the French armed resistance. A number of pioneers in hiding in the greater Paris area had joined this.

Max Windmüller, who had succeeded Reilinger, had allowed 'headquarters' employees who were not arrested (such as Paula Kaufmann, Hans Ehrlich and Metta Lande) to lend a hand to this resistance. In one instance, Kaufmann had taken a job as a secretary with the *SiPo/SD*, where she could get at all kinds of information, including about German defence works in Paris. Fritzi Okladek, an Austrian pioneer, worked at a German printing business, where she could steal all sorts of documents. Metta Lande, Windmüller's girlfriend, worked as a courier.[27]

Josef 'Ad' Linnewiel was the liaison with this French resistance. Linnewiel was a Jewish student born in Arnhem in 1920, who fled on his own to France in 1943. He had intended to go Spain, but this failed. However, Linnewiel became involved in the transport of Allied pilots to Spain, via 'King Kong' Lindemans and others. In this way he came into contact with the *AJ* and the Palestine pioneers.

Linnewiel was part of a Jewish resistance group in Paris, together with pioneers such as Ernst Asscher, Paul Wolff, Leo Weill, Ludwig Jakobs and Gert Sperber and his wife Anna Sperber-Chlebowski. This group was led by Ernest Appenzeller, who came from Nice, where he had participated in armed initiatives of the *AJ*. Appenzeller's Paris group was part of the umbrella *Forces Françaises de*

---

25 Lazare, *La résistance*, 304, 305. See also Latour, *La Résistance*, 195, 196.
26 GFHA, Cat. no. 47, Verklaring Lolly Eckhard, 4.
27 GFHA, Cat. no. 108, Interview Kurt Benjamin-Paula Kaufmann, 3, 4 (Translation from Hebrew to Dutch by Ithamar Lehmann). See also GFHA, Cat. no. 47, Verklaring Lolly Eckhardt, 4.

l'Intérieur (FFI, French Forces of the Interior) and was involved in weapon transport, sabotage and raids on German institutions and the liquidation of traitors.[28]

Through his contact with Porel, Max Windmüller had seen a possibility of communicating with Kurt Reilinger, who was in prison. On 11 July 1944, he gave 'Lydia' Tcherwinsky a letter for Reilinger with questions about his arrest and possibilities for an escape. Two days later, Reilinger was taken to the SD headquarters in the Rue des Saussaies to write an answer to the letter that the Germans had prepared.[29]

On 17 July, Lazarus and Kapel were to leave for London, via Tours. A car stood ready at the agreed place, but the two men were overpowered and taken directly to the infamous SD office in the Rue de la Pompe. The next day, a new step in the cooperation between Porel and employees and the *AJ* was to be discussed in a hotel room in the Rue Erlanger. The meeting was raided by about fifteen agents, who arrested five key people of the *AJ*, including Appenzeller and Pohorylès. Max Windmüller, who was also there, was arrested as well.

In the following days, there were twelve more arrests, which included Paula Kaufmann, Ernst Asscher, Gert and Anna Sperber and Paul Wolff. They were all locked up in the Rue de la Pompe. Ad Linnewiel and Metta Lande escaped by coincidence. Hans Ehrlich, who had avoided contact with Porel because he did not believe his stories, tried to warn several of his comrades. The German police, lying in wait for him at a hotel, shot him in the leg when he tried to escape.[30]

After heavy interrogation with serious mistreatment, whereby one French resistance man died, the arrestees were taken to the Drancy detention camp. With the information obtained, for which several agents had also come over from the Netherlands and for which Joop Westerweel was probably also interrogated in Vught, the Germans were able to form a good picture of the organization and the way it worked. Willy Hirsch and Zippi Fränkel were locked up as 'heavy cases' in the Drancy prison.[31]

---

28 Latour, *La Résistance*, 183, 184. See also CDJC Paris, DLXI-61, Verklaring Ad Linnewiel 1–18, 1974. The statement contains several inaccuracies and is not very reliable. GFHA, Cat. no. 65, Verklaring Ernst Asscher, 5. The *Armée Juive*, called itself *Organisation Juive de Combat* from mid-1944. In this book, however, the name *Armée Juive* is used.
29 GFHA, Cat. no. 108, Verklaring Paula Kaufmann about the 'Affaire Lydia Tschervinsky – Charles Paul', 1946.
30 GFHA, Cat. no. 48, Verklaring Hans Ehrlich, 2. See also CDJC Paris, Verklaring Jeanne Latchiver, DLXI-55.
31 GFHA, Cat. no. 55, Verklaring Alfred Fränkel, 6.

Transport to Germany

At the beginning of August 1944, the Allies were already close to Paris. The Drancy commander, SS officer Alois Brunner (a close collaborator of Adolf Eichmann), decided around the 15th of August to flee with his staff, by train to Germany. A wagon was coupled to the transport that carried 51 Jewish prisoners. What Brunner had planned to do with them is not clear. He may have wanted to put them in a show trial against Jewish 'terrorists' or to use them as hostages.

The passengers in the last wagon were almost all involved in the Porel affair. However, several Jewish dignitaries such as the airplane designer Marcel Bloch (Dassault), a Russian princess of Jewish descent and the director of the Rothschild Hospital in Paris with his family had also been added. Passengers who belonged to the Westerweel group included Kurt Reilinger, Max Windmüller, Zippi Fränkel, Paula Kaufmann and Paul Wolff. Suzi Hermann and Lolly Eckhardt were able to stay in Drancy and were released shortly afterwards, possibly by mistake. Anna Sperber, whose Jewish identity the Germans could not establish, had previously been sent as a political prisoner to the women's camp Ravensbrück, where she died. Hans Ehrlich, who also kept to his false identity, was sent to Buchenwald as a non-Jew. He escaped during a transport at the end of the war.

The French resistance people took tools baked in large loaves of bread from Drancy. During the journey, which went very slowly as a result of sabotage and bombardments, there was an opportunity to break the wagon open. It was agreed that they would jump from the train during the night journey in groups of two—each time one Frenchman and one Dutchman. This agreement was not kept.

To be able to get some rest, they all had agreed that everyone should first go to sleep, after which the escape could start. However, when the Dutchmen woke up, it appeared that all the French resistance members (including Jacques Lazarus, Ernest Appenzeller, René Kapel and Henry Pohorylès) plus several other passengers (including Marcel Bloch) had disappeared—a total of 26 men. Of the Dutch group, only Ludwig Jakobs, who had little contact with the other pioneers, had escaped. 'There was probably a break in the communication line somewhere', Pohorylès told the Israeli-French historian Renée Poznanski in 1987.[32] This explanation appears, however, to be far too simple. During the im-

---

[32] Renée Poznanski, *Jews in France during World War II* (Hanover, NH: University Press of New England, 2001), 572, note 86. Several statements were made about what happened in 'the last wagon' and even a novel *Le dernier wagon*, (J-F Chaigneau, Paris: Julliard, 1981) was published.

prisonment, a lack of trust had probably arisen between the French and Dutch resistance members.

When the Dutchmen wanted to jump out of the train, it stopped. 'If the train had only gone on for another quarter of an hour, the whole wagon would have been empty!', Kurt Reilinger wrote in his post-war report.[33] The German guards opened the wagon and saw what had happened. The remaining prisoners had to get out and would be shot dead. They had to dig a large hole, but there was disagreement among the German guards. An officer of the air force, who apparently did not want to be involved in the execution, intervened and the prisoners were allowed to get back in the train.[34]

After all sorts of delays, they arrived at the Buchenwald concentration camp at the beginning of September 1944. The Germans sent Paula Kaufmann and the other female prisoners through to Auschwitz, via the prison of Weimar. In Auschwitz, Paula was admitted as a non-Jewish prisoner on the basis of her false papers. When the camp was evacuated at the beginning of 1945, she was transported to Bergen-Belsen, where she was liberated.

The remaining members of the Paris group quickly lost sight of each other in Buchenwald because they were sent to different outside camps. Max Windmüller went to a camp near Bochum shortly after arrival. He worked there until March 1945 in very bad circumstances at a steel company. Paul Wolff had to go to a satellite camp in Mühlhausen.

The Germans sent Kurt Reilinger, Ernst Asscher and Gert Sperber to the infamous Dora-Mittelbau camp in the Harz Mountains, where rockets were made in underground tunnels. Sperber died there, probably in a bombardment. Asscher went via several other camps to Bergen-Belsen, where the British liberated him. Willy Hirsch also ended up in Bergen-Belsen. However, he was so weakened by then that he died shortly before the liberation. Kurt Reilinger was placed in a construction brigade, which had to repair damaged railway lines. He was finally liberated in Flensburg by the English army—and then went to Sweden for several months to recover.

Windmüller and Wolff returned to Buchenwald in March 1945. When the American troops approached the camp at the beginning of April 1945, the camp commander transported more than 3,000 Jewish prisoners to the Flossen-

---

However, none of the statements give any clear idea of the motives of the French for their actions.
33 GFHA, Cat. no. 45, Lore Durlacher, Brief Kurt Reilinger aan Ilse Birnbaum, Lore Durlacher, Frans Gerritsen and others, 15 mei 1945.
34 GFHA, Verklaring Fränkel, 6. Metta Lande and Ad Linnewiel and several French resistance fighters tried to follow the train in a car. However, they lost the transport in Luik (Liège).

bürg concentration camp, close to the Czech border. Later, some of the other prisoners followed. The journey was by foot and by train in open freight cars. The primary objective of these transports was to kill as many prisoners as possible.

After a short stay in Flossenbürg, the Germans also cleared out this camp. The 22,000 prisoners, including Paul Wolff and Max Windmüller, had to walk to Dachau. On the second day of the march, a guard shot Max Windmüller dead because he could not walk fast enough. Paul Wolff, who witnessed this, was liberated the next day by the Americans. Zippie Fränkel, fearing further deportations, had escaped from Buchenwald shortly before the liberation and was hiding in the woods. The Americans quickly hospitalized the severely weakened Fränkel.[35]

Courage and Overconfidence

The French episode was characteristic for the courage, perseverance, inventiveness and feelings of camaraderie of the Palestine pioneers, most of whom were 25 years old and younger. With only limited support from their Dutch mentors and through their work with the *Organisation Todt*, they managed to build up a properly functioning illegal network, make contacts with the Jewish resistance in France and organize the crossings to Spain. Moreover, several *chalutzim* contributed to the French armed resistance. That approximately 70 pioneers arrived safely in Spain was a noticeably good result. About the same number survived by hiding in France.

However, these characteristics of the *chalutzim*, in combination with the idea of being free in France and more or less invulnerable, also led to feelings of self-overestimation and a misunderstanding of the German capacity to retaliate. To his amazement, the French *Maquisard*, Marc Levy—who fought with several pioneers for some time—saw them go into a railway wagon reserved for the German army. 'Look out! You are risking prison', he called, but the Dutchmen laughed at him and produced their German travel documents.[36]

The journeys taken using the German travel documents were copied by the French-Jewish resistance on a modest scale. The same cannot be said for visits to German military canteens and cinemas, and stays in hotels reserved for Ger-

---

35 Meyer-Dettum, *Max Windmüller*, 31. See also GFHA, Verklaring Fränkel, 6. GFHA, Verklaring Paula Kaufmann about the 'Affaire Lydia Tschervinsky – Charles Paul', 1946. GFHA, Cat. no. 266, Verklaring Paul Wolff, 24 april 1989.
36 Quoted in: Latour, *La Résistance*, 108.

mans: these activities were regarded by French resistance members with a combination of admiration and wariness. The very risky practice of travelling to the Netherlands on leave passes was mentioned earlier. Menachem Pinkhof's warning after his arrest to focus on the work of hiding and saving Jews—and not to engage in dangerous adventures—had clearly not reached France.

In the spring and summer of 1944, the two waves of arrest that put a large part of the Paris staff and armed resistance members in German cells (and later in detention camps) were partly the result of considerable carelessness.

Joop Andriesse's journey to Switzerland appears to have been poorly prepared. His arrest offered the Germans the opportunity of arresting Kurt Reilinger and his closest assistants. The mixing of hiding and active resistance activities after this time made the Palestine pioneers vulnerable for German counter measures. The Porel affair and Max Windmüller's attempt to communicate with Kurt Reilinger through Porel were admittedly an example of camaraderie, but they were also careless, with fatal consequences for Windmüller himself and several other pioneers.

12 After the Liberation

It is now largely known how an end came to the Westerweel group's work, both in the Netherlands and in France, in the summer and autumn of 1944. With the German capitulation at the beginning of May 1945, about a 100 of the Palestine pioneers connected with the Westerweel group were still in hiding in the Netherlands. Others had stayed in France and approximately 70 had gone to Palestine via Spain. Dozens of pioneers were detained in German camps. Most of the Dutch core members were also in concentration camps; others had gone into hiding.

This chapter first takes a brief look at the further activities of several members of the group, starting in the autumn of 1944 and continuing as long these activities were linked to the *Hechalutz* and Palestine.

Next, a look is taken at how the resistance activities, such as those undertaken by the Westerweel group and other resistance groups, affected the chances of survival of the Palestine pioneers.

The final part of this chapter focuses on how the Westerweel group's work was evaluated and remembered after the war, both in the Netherlands and in Israel.

Hechalutz Work and Aliyah

The Netherlands was liberated in phases. In the autumn of 1944, the Allies reached the large rivers, the Maas (or Meuse) and the Waal, liberating the southern provinces. The liberation of the northern provinces and large parts of Gelderland and Overijssel followed in the early spring. At the beginning of May 1945, the German army capitulated in the west of the country.

Immediately after the liberation of Sevenum in November 1944, Lily Kettner started to look for Jewish children hiding in northern Limburg. At first, Lily went out every day on a bicycle without tires to ask all of the pastors in the region where Jewish children were staying. She later visited them again and organized meetings.

When she had established contacts throughout Limburg, and in Eindhoven with Jewish circles, Lily and her helpers started publishing a stenciled newspaper once every two to three weeks. The newspaper contained 'stories about Israel and the kibbutz, and so forth.', plus riddles and songs that they taught the children. Lily was given a typewriter and paper for this purpose by Jewish soldiers in the American and British army. At Passover, they also managed to obtain matzos (unleavened bread). The children remembered the matzos they had before the war, but said that these had been round, and not square like the American matzos. The municipality of Heerlen arranged a venue for the meetings.[1]

Lily and her helpers were not welcome everywhere. At one address, the lady of the house, who had taken in a Jewish child, said: 'We have raised him religiously and you are not now going to confuse him'. The door was then shut in Lily's face. But mostly, the reception she received was friendly and people would say: 'Look what this lady has brought you for Passover.' Also, a leader of a Catholic youth organization who lived in Tienray offered her house to celebrate Purim.

It was often not easy to remind the children of their Jewish background. Most of them had become attached to their largely Catholic foster parents and, quite understandably, sometimes did not want to leave them. One girl originally wore a cross and did not want to take it off. Later, she hid it under her clothing, but finally, after a month, took it off.

Lily Kettner would have loved to open a Jewish children's home immediately in November 1944, but she could not find the necessary support and funding for this cause. She was not at all impressed by the Jewish organizations, which she considered ineffective.

After May 1945, the work of the Limburg group was taken over by Harry Asscher and his wife Hanna. They brought children whose parents had not returned from the camps to a home in Dieren. Most of these children went to Palestine at a later date.[2]

In June 1945, Menachem Pinkhof and Mirjam Waterman returned to the Netherlands from Germany. Together with soldiers of the *Jewish Brigade* who belonged to the British army, and Dutch Zionists, they immediately started reorganizing the *Hechalutz* work. A stenciled newspaper, the *Dawar Hechalutz*, was published and several meetings were held. However, Menachem was forced to stop and rest for six months after he was diagnosed with a lung complaint that he had contracted in Bergen-Belsen. Shortly after his recovery, in April 1946, he left for Palestine with Mirjam, to whom he was now married.[3]

The *hachshara* did not undertake any more major activities. A few training centers were started, in the Catharinahoeve in Gouda, for example. However, most pioneers quickly left for Palestine. Before they left, several of them attended a supplementary agriculture course. Chaja Waterman, for example, who hid in Friesland during the war, specialized in dairy production.

---

1 GFHA, Cat. no. 111, Verklaring Lily Kettner, 5, 6, maart 1957.
2 Ibid. See also Interview Mirjam Pinkhof-Lily Kettner, 1–6, 1987.
3 GFHA, Cat. no. 173, ongedateerde Verklaring Menachem Pinkhof, 1–12.

At the end of August 1945, Kurt Reilinger also returned to the Netherlands, from Sweden. He threw himself into the *Hechalutz* work as before and went to work on *Dawar Hechalutz*. Fourteen days after his return, when he was hitchhiking one evening from Amsterdam to Groningen to get his papers organized, he was run over by an army truck. Only one headlamp of the car was lit and Kurt probably thought that it was a motorcycle. Unfortunately, he stepped too far out into the road and died on the spot.

His death was a heavy loss for the remaining Palestine pioneers, who printed a memorial of him in their newspaper. Kurt Reilinger represented to them an important part of their collective past from the years of persecution. Menachem Pinkhof wrote:

> Deserted Ghetto of Amsterdam. Bare hotel rooms in Paris.
> Full night trains to dark cities.
> The care for young peoples' lives beat in our hands.
> The fires of Auschwitz lit in the night, and we believed in life'.[4]

In his letters from Sweden, Kurt Reilinger had stated that he planned to stay in the Netherlands only for a short while. Once his papers were in order, he planned to travel on to Palestine. Although Zionism remained his ideal, he also wrote that he had been impressed by the sacrificial attitude of, and the close bonds of togetherness he had shared with, other non-Jewish prisoners in Buchenwald.

The Illegal Aliyah

Most of the *chalutzim* group that had been in hiding in the Netherlands as well as most of the prisoners from Germany went to Palestine after the war. Before going, almost all of the ex-prisoners returned to the Netherlands first in order to get their papers arranged and to meet the other pioneers. They then travelled through to France. The reception in the Netherlands was, however, sometimes far from friendly. For example, Zippi Fränkel was imprisoned for some time after his return from Buchenwald, because his identity could not be properly determined. This was partly due to the false papers that he had used.[5]

In August 1945, Lily Kettner's group joined a transport of the *Jewish Brigade*. The unit's soldiers had become involved in Operation *Berisha* (new beginning) after combat actions in Italy and Germany. Several Jewish resistance leaders in

---

4 *Dawar Hechalutz*, oktober 1945, 1.
5 GFHA, Cat. no. 55, Verklaring Zippi Fränkel, 6.

Eastern Europe had concluded during the war that Jewish life had been destroyed there. It therefore made no sense to try to rebuild it, considering the endemic anti-Semitism in the region. Operation *Berisha* would take Eastern and Central European survivors of persecution to Palestine.

To transport the displaced persons from the refugee camps across the whole of Europe and the Mediterranean Sea, the *Jewish Brigade* and other organizations that were involved in the operation used false papers. With these papers, fake army units were set up to confuse the British and other authorities. The operation was financed by the *Joint*.[6]

The *Brigade* soldiers gave Lily and the other pioneers some better clothing, 'a pair of trousers for one, a coat for the other', and took them to Brussels and on to Paris. From Paris, Lily and her travelling companions took the train to Marseille. After some wandering about, they ended up on the other side of the Pyrenees near Pau, in a shelter that was run by Lolly Eckhardt and Zippi Fränkel. Here, Lily met other Palestine pioneers, including Metta Lande, with whom she travelled back to a camp near Marseille.

From Marseille, buses took the pioneers to the New Zealand ship the *Mataroa*, which was in the harbor. Some of the passengers, including Metta, had legal papers. Others who did not have these papers travelled as stowaways in the ship's hold. After arrival in Haifa, they were smuggled to land and offered sanctuary in a kibbutz.[7]

Two Dutch members of the Westerweel group, Frans Gerritsen and Jan Smit, also became involved in the activities of the *Jewish Brigade*. They had had the foresight to make plans earlier for this purpose. On 5 or 6 May 1945, Jan and Frans had held a discussion in the Amsterdamse Bos with Lore Durlacher, Tinus Schabbing, Henny Gerritsen and several others. There, they had founded the JAZ, the *Jüdische Auswanderung für Zionisten* (Jewish emigration for Zionists). The purpose of the JAZ was to smuggle Palestine pioneers with false papers to France, more or less as was done during the war.

'Do-er' Frans started immediately. He bought a small truck, probably with money from the *Hechalutz*, and drove it to Paris. He purchased Allied army uniforms in an army surplus shop. He then packed the uniforms and took them back to the Netherlands in large duffle bags. In an American army camp near Enschede, a quartermaster sergeant gave him shoes, socks, coats and uniforms.

---

[6] Mark Wyman, *DPs: Europe's Displaced Persons, 1945–1951* (Philadelphia: Balch Institute Press, 1998), 131–155.
[7] GFHA Cat. no. 111, Interview Mirjam Pinkhof-Lily Kettner, 4, 5.

However, Frans's expedition did not continue. In the summer of 1945 he was told that it would be better to work with the *Jewish Brigade*, which was doing similar things. At the beginning of September, he was with a group of children in a barn at the Waterman family's house in Loosdrecht for the first transport with this *Brigade* to France.

> Then someone came in and asked if there was anybody who knew about cars. I asked: is something wrong? He said: 'Heaven help us, you're still alive Frans!', it was Kurt Reilinger—that's how things went with the Brigade. You just sat in the back of the car in uniform. When you got to the border, you saw lots of soldiers and greeted them. In France, there were different reception centers and hachsjaras.[8]

Wil Westerweel also visited France at the end of 1945. She travelled to Marseille and the reception center at Pau, also dressed in uniform and with an identity card from the French resistance. In both camps, she met many old acquaintances, including some from Loosdrecht—which particularly did her a lot of good. On the way back, it appeared that her camouflage was far from perfect. When Wil was eating in a military canteen with several travelling companions, an English officer came up to them and said 'that he knew that we were not soldiers'. However, because it was just before Christmas, he did not want to cause any problems.[9] The *Jewish Brigade* was dismantled in the summer of 1946.

In 1947, Frans and Henny Gerritsen stayed in Palestine for several months to see if they could settle there. During this time, Frans also acted as advisor in *kibbutzim* regarding preparing secret weapons caches. These caches were his specialty. The pacifist Gerritsen, who had never wanted to carry a weapon during the war, now became involved in military matters. He thought that the realization of a Jewish state was a necessity, considering the persecutions during the Second World War. 'If you do not have your own country somewhere, then that will be the end. It's easy for me, I have a country. I have the Netherlands'. This was his reasoning, he told Yigal Benjamin more than forty years later.[10]

In his cooperation with the Palestine pioneers, Jan Smit went a step further. His Jewish wife, Helga Decker (a pioneer who came from Breslau) and thirty other *chalutzim* sailed with the *Tel Chai* to Palestine in March 1946. The intense contact he had experienced with the people he had cared for—almost all of whom were his own age—had led to a feeling of friendship and kinship. Jan Smit followed his wife a short time later. One of the reasons for Jan going to Pal-

---

8 GFHA, Cat. no. 63, Interview Yigal Benjamin-Frans Gerritsen, 19 december 1989, 1–18.
9 Westerweel, Lijn of cirkel, 102, 103.
10 Ibid. See also GFHA, Cat. no. 63, Interview Yigal Benjamin-Frans Gerritsen, 18.

estine was that he was not attracted by the thought of returning to civil society after the turbulence of the occupation. However, he did not want to become Jewish and he was also not a real Zionist. He considered Zionism to be too much a form of nationalism, which was difficult to mesh with his left-wing socialist views. These views caused tensions with others, and Jan and Helga returned to the Netherlands after three years. His relations with the Westerweel group members living in Israel, however, remained excellent.[11]

Taking Stock

Only after the war could a proper evaluation be made of how the persecution of the Jews affected the Palestine pioneers. Two overviews provide some idea of which problems encountered in taking stock.

The first overview, compiled by Ineke Brasz, was included in the book about the *Jeugdalijah* in Loosdrecht. This book starts with the situation at the beginning of the war, when the total number of pioneers was 821.[12]

The second overview, compiled by Yigal Benjamin, starts with the beginning of the persecution of the Jews in July 1942. According to his calculation, the number of pioneers was then 716 (he did not count the 59 pioneers from the *Werkdorp* who were deported to Mauthausen in June 1941.) According to Benjamin, the number of mostly Dutch religious pioneers in places like Franeker and Laag-Keppel in July 1942 was more than 40 lower than the numbers estimated by Brasz. The explanation for this is that it is not clear where these pioneers went after the *hachshara* in Beverwijk was closed in the beginning of the occupation. Several dozen probably returned to their families; the exact number is not known.

As a result of the calculation methods used, Benjamin's estimate of the number of surviving pioneers is more than 10% higher (54.9%) than the estimate of Brasz (44%). However, both overviews feature unavoidable and mostly limited deviations for the different categories. These deviations are mainly due to the continual changes in the size and compositions of the different groups that occurred during the war. The results shown here must therefore be viewed with some caution.

This is partly why only Benjamin's overview is included and used in this research, as it gives a better and clearer picture of the results of the hiding operation for the different pioneer groups.

---

11 GFHA, Cat. no. 225, Interview Yigal Benjamin-Jan Smit, 1–10, 18 april 1990.
12 Benjamin, ibid., 22, 23. See also Verklaringen Pinkhof, 14.

Apart from the case of the Elden pioneers, going into hiding always appeared to be better than deportation. As described in the case of Elden (with 72.1% survivors), the possession of Palestine certificates and the protection of Manfred Samson's father were important assets. The members of the Elden group were deported in 1944, but mainly to Bergen-Belsen, where the chances of survival were better.

Tab. 1: *Hachshara* members July 1942-May 1945, survivors and victims

Hachshara	Total	Survivors		Deceased	
		Number	Percentage	Number	Percentage
Deventer	268	155	54.9	113	42.2
Wieringermeer	189	91	48.1	98	51.9
Laag-Keppel	34	8	23.5	26	76.5
Beverwijk, Franeker	52	34	65.4	18	34.6
Loosdrecht	51	33	64.7	18	35.3
Gouda	27	18	66.7	9	33.3
Elden	43	31	72.1	12	27.9
Agudah Enschede	52	23	44.2	29	55.8
Totals	**716**	**393**	**54.9**	**323**	**45.1**

The *hachshara* group's members in Westerbork who did not come from Elden were also able to profit from this protection or from the Palestine certificates—be these legal or falsified. A place in this group provided a respite from deportation.

From the beginning of 1944, also the pioneers who did not belong to the Elden group were mostly deported to a more favorable camp, such as Bergen-Belsen or Theresienstad. Several (about 25) were able to escape. It must be noted, however, that members of the *hachshara* group were used to fill transports to Poland on several occasions, when there were not enough people in the train to meet the quota. The protection was therefore certainly not watertight.

The Westerweel group helped the entire Loosdrecht and Gouda *hachshara* groups with hiding. They also helped individual pioneers from the *Werkdorp*, the Deventer group and, to a lesser degree, people from the *Mizrachi* groups in Beverwijk and Franeker. The Westerweel group cared for a total of about 300 pioneers.

With regard to people from Loosdrecht and Gouda, roughly equal percentages of 64.7 and 66.7%, respectively, survived. For the pioneers from the *Werkdorp* and Deventer, the percentages were much lower. For Beverwijk and Franeker, the percentage was almost equal to that of Loosdrecht and Gouda—but most of these pioneers were not helped by the Westerweel group.

The lowest percentage of survivors came from the religious *hachshara* of Laag-Keppel. As previously mentioned, the leaders there opposed going into hiding for religious reasons. In April 1943, the group of 34 people was sent to Vught; they were deported almost immediately via Westerbork to Auschwitz.

The leaders of the ultra-religious group living in isolation in Enschede acted differently. They supported the hiding of individual members and, helped by the Jewish Council in Twente and the reformed minister Leen Overduin, found a hiding place for almost half of the pioneers. In several cases, the help also went through the Westerweel group.[13]

The Joop Westerweel Forest and Beyond

Most surviving pioneers settled in Palestine in *kibbutzim* of various movements within the socialist mainstream. A large group chose kibbutz Gal Ed, about 30 kilometers south of Haifa. Others settled in Hazorea, and from there some of them founded kibbutz Yakum near Netanya in 1948. This kibbutz belonged to the left-wing socialist movement *Hashomer Hatzair*. Several others chose a religious kibbutz or went to live with family elsewhere.[14]

The Palestine pioneers had great appreciation for Joop Westerweel, the man who, together with others, had organized the hiding operations in Loosdrecht and who had also played an important part in other resistance activities, such as developing the escape route to France. When Joop was arrested in March 1944, Kurt Reilinger immediately travelled to the Netherlands to see if he could help secure his release. The news of Joop's execution, five months later, made a deep impression on them.

In 1947, the *chalutzim* therefore took the initiative to start planting the Joop Westerweel Forest, together with the Jewish National Fund in kibbutz Dalia, close to Gal Ed. In 1954, the Forest had grown to the point that a modest monument could be placed with the following text on a grey natural stone plaquette.[15]

---

13 Benjamin, ibid., 24–28.
14 Asscher, *Van Mokum*, 225–228. See also Siegel, *Locomotieven*, 186–207.
15 https//:www.4en5mei.nl, joopwesterweelwoud, accessed 20 september 2014.

Ter herinnering aan	To the memory of
Joop Westerweel	Joop Westerweel
Die bezielende kracht was in	Who was the inspiring force in
het ondergrondse werk tijdens	the resistance work during
de Duitse bezetting van Nederland	the German occupation of the Netherlands
zijn leven gaf voor de redding	and gave his life to save
van de Joodse jeugd	Jewish youth

Ten years later, the former Palestine pioneers and the Israeli government organized a reception to honor the members of what was then beginning to be referred to as the Westerweel group. During a visit to the memorial center of Yad Vashem, 15 members received an award from this institute. They were also inscribed in the register of the *Righteous Among the Nations.* Yad Vashem had only begun to give these awards the year before, in 1963. The group was also received by President Zalman Shazar and Minister of Foreign Affairs, Golda Meir. This was therefore a high-level, official appreciation for the Westerweel group's work.

The relationships between the former Palestine pioneers and the Dutch members of the group remained excellent, and a second generation gradually started to play an important part. For a long time, the key figure in the contacts

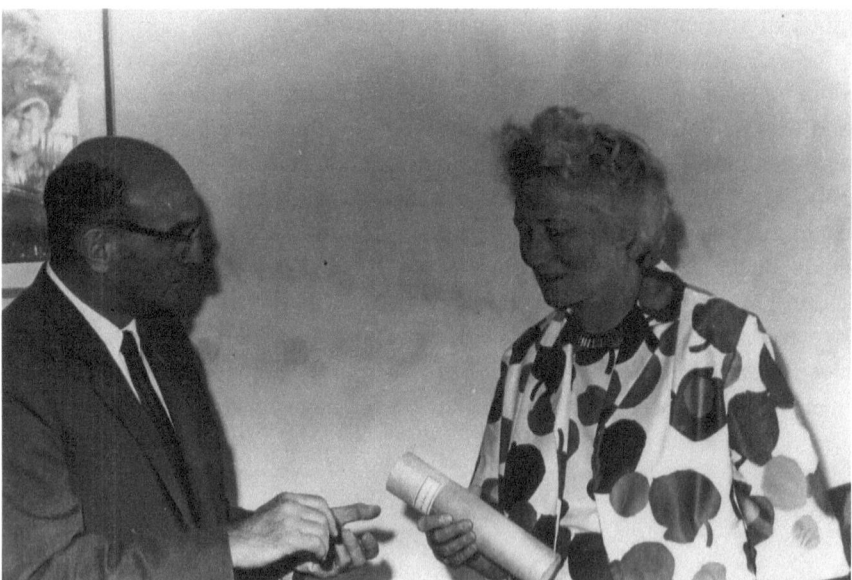

Fig. 29: In 1964, Wil Westerweel received the Yad Vashem award as 'Righteous among the nations' from Gideon Hausner, the prosecutor in the Eichmann trial and chairman of the board of Yad Vashem.

between the groups was Wil Westerweel, who regularly stayed in Israel—partly because her daughter Marta was married to an Israeli. Wil had also learned Hebrew and therefore often acted as an interpreter and guide during visits by other members of the group.

In the 1970s and 1980s, however, Wil became increasingly dissatisfied with the political developments in Israel, especially the occupation of Palestinian areas. In speeches at remembrances, she always tried to connect the past and present, 'and to make it clear that it only made sense to commemorate what took place from 1940 to 1945 if our behavior now is a consequence of it'.

Wil did not want to judge the Israeli members of the group, who had lost many family members and who, it seemed to her, had to stand by and watch how their country had developed in a completely different direction than they probably had foreseen and hoped. However, she experienced the 'eternal remembrance' of murdered family members as increasingly promoting an agenda which prevented a discussion 'about what we can do now to change today's untenable situation'. According to Wil, the occupation politics of Israel had maneuvered her former comrades-in-arms against their will into the role of oppressors.

Wil had explained her view to several friends in Israel, 'but I fear that that it is not understood. And I find that a pity, because no matter how unpleasant I find it to hurt them, I cannot be any different than I now am'.[16]

Her critical attitude, however, did no harm to the good relations with the Palestine pioneers. She and the other Dutch members of the Westerweel group continued to visit Israel regularly.

In the Netherlands, there was less official recognition for the Westerweel group. To the best of my knowledge, none of the members received an award. However, professors L. de Jong and J. Presser mention the group in their historiography. Attention is also paid to the group in the Amsterdam *Verzetsmuseum* (Resistance Museum) and in the permanent display in the National Monument at Camp Vught.

In several places, streets were named after Joop Westerweel. In the Amsterdam working-class district of De Baarsjes, also a primary school bears his name.

---

16 Westerweel, Lijn of cirkel, 128, 141, 142.

13 Helpers and Non-conformists

The previous chapters describe the history of the Westerweel group, in which Jews and non-Jews worked together to take the largest possible number of Palestine pioneers to safety, away from persecution. Initially, the group members improvised, but later they were better organized. Joop Westerweel and other members first took the Loosdrecht *Jeugdalijah*, and then as many other Palestine pioneers as possible, to safety in the Netherlands. Then, from the beginning of 1943 until the spring of 1944, some 150 of the pioneers were transported safely to France; from there, 70 went on to Spain.

Despite setbacks, these operations were reasonably successful. The group managed to keep more than 250 Palestine pioneers out of the hands of the Germans. However, as we've seen, the previous few chapters clearly show the consequences of this resistance. The Sipo-SD largely dismantled the Westerweel group, first in the Netherlands and then in France—with all the associated consequences for the arrested members.

This last chapter attempts to answer several central questions formulated in the Introduction around the history of the Westerweel group. A complicating factor was that this group consisted of both Jews and non-Jews. The Jews were under immediate threat, but the non-Jews had various options from which to choose.

The Jews had, in fact, no other possibilities than to either follow or resist the anti-Semitic directives of the Nazis. The anti-Jewish measures, of course, did not affect the groups' non-Jewish members, who had several alternatives to get through this difficult period. Roughly speaking, the following possibilities were open to them:
- They could support the Nazis to a lesser or greater degree in carrying out the measures, which is what the members of right-wing radical organizations did.
- They could watch powerlessly or act as if the persecution of the Jews was none of their business.
- They could adapt themselves to some of the conditions, as the Nederlandse Unie and the College of Secretaries-General did, and focus their protest against the worst of the persecution.
- They could choose resistance against the anti-Semitic measures.

This chapter aims to formulate an answer to the central question arising from this last possibility: Why does someone help others, with no material or other advantage for him- or herself? And, regarding the subject of this book: What motivated the Westerweel group's members to offer help to their Jewish fellow-citizens?

A second question is: How can the Westerweel group's initiatives be characterized from a social perspective and how did the group differ from, or to what extent was it similar to, other resistance organizations?

The Psychology of the Helpers

Much research has been done during the past decades to find an answer to the first question, with special attention being paid to the persecution of the Jews during the Second World War. Why did non-Jews risk grave danger to help Jews during this war? In the occupied Eastern European countries, people who helped Jews ran a great risk of being immediately killed if the Germans caught them. Yet, there were many men and women in Eastern Europe who were ready to help Jews—as recorded, for example, in the documentation of *Yad Vashem*.

In Western Europe, the Germans acted less brutally. In the Netherlands, the Germans tried to avoid resistance from the Dutch population by initially often leaving alone the 'Jew helpers' (as they were called). A frequent explanation was that the helpers had been misled by the Jews. Sometimes, though, they were punished with six weeks detention. Later, the punishment was increased to a six-month prison sentence in the Vught or Amersfoort camps. From 1944, there was the risk of deportation to a concentration camp in Germany, which was perilous. During the last nine months of the war, 'Jew helpers' were sure to be detained. In several cases, the Germans executed them as a reprisal for acts of resistance, just as they executed other political prisoners. This was Joop Westerweel's fate.

Samuel (Shmulek) Oliner is one of the earliest researchers into the motives of people who helped Jews. Oliner was a Jewish boy of twelve when, in 1942, the Polish ghetto where he lived was brutally cleared out by German soldiers and their Ukrainian helpers. Because he had gone outside the ghetto earlier, to look for food, and knew his way around, his step-mother sent him away and told him to save his own life. Barefoot, and wearing hardly any clothes, Samuel fled to the house of Balwina, a peasant woman whom he had previously met with his father, and who had been friendly to him.

Without thinking about the danger, the woman took him into her house. Because he could be recognized by the neighbors, she hid him for some time in the basement. Then she made sure that he got another name, Jusek Polewski, and a Catholic identity. After this, she sent him to herd cows in a nearby village, for a childless married couple. An older son from the Polish family visited him now and then to see how he was getting on and to tell him the news. This is how Shmulek/Jusek kept alive and survived to see the liberation.[1]

After the war Samuel Oliner went to the United States, where he studied social sciences. During his studies, he became increasingly fascinated by what had happened to him during the war. He decided to make the behavior of Balwina and other helpers the subject of his research. In 1988, together with his wife Pearl, he published a pioneering study about 700 people who had risked their lives to help Jews and non-Jews, or had been helped themselves, during the Second World War. The Oliners' most important conclusion was that something like an altruistic personality does exist, and people with this personality had been ready to help Jews—mostly without much deliberation.[2]

The original results of the Oliners' research are often criticized because they were based on statements that were sometimes made long after the war. Consequently, they could give a distorted picture of the situations and the people involved. Later research gives a more nuanced picture of this type of personality. In his 1997 dissertation about the persecution of Jews in the Netherlands, the British researcher Bob Moore takes a publication by the American social historian (and supporter of the altruism theory) Lawrence Baron as his starting point. In this publication, Baron describes four types of people who could have become involved in helping Jews:

1) People with a positive attitude to Jews, based on earlier experiences.
2) People who had feelings of affection or friendship for Jews, or somehow felt linked to them.
3) People who helped out of compassion and empathy, who were inspired by their own experiences of persecution or feelings of being socially marginalized.
4) People who were driven by moral (ideological and religious) considerations.[3]

Baron's subdivisions are distinct, because he names specific groups of potential helpers. The borders between the four categories are, however, very vague. Moreover, he does not address why one person from a certain group does help, and another does not.

---

[1] Samuel and Pearl Oliner, *The Altruistic Personality: Rescuers of Jews in Nazi Europe* (New York: Free Press, 1988), xv-xvii.
[2] Oliner, *The Altruistic Personality*, mainly 142–170.
[3] Bob Moore, *Victims and Survivors. The Nazi Persecution of the Jews in the Netherlands* (London: Arnold 1997), 177–179. See also Lawrence Baron, "Dynamics of Decency," in *The Nazi Holocaust, part 5, volume 2*, ed. Michael Marrus (Munich: Pieper, 1998), 608–628.

Other authors have highlighted the important role played by formal and informal leaders in a community. One example concerns two French ministers in some small, mainly Protestant villages in southern France, where hundreds of Jews were hidden. In Polish convents, the Mother Superior determined whether Jews would be hidden. Moore himself mentions the role of Dutch Reformed ministers in the resistance and in helping Jews.[4]

In the Westerweel group, Pastor Vullinghs was a telling example of the influence of such a leader. From the political left wing, in addition to Joop Westerweel and his helpers from the *Werkplaats*, the leader of the Almelo 'free socialists,' Derk Senechal, deserves mention.

The American historian and psychoanalyst Eva Fogelman created a typology comparable to Baron's subdivisions. However, she also pointed out the importance of 'enabling situations', such as a request for help, the presence of a place to hide, the existence of an organization, and mainly the luck of the right moment at which a call was made to the relevant person.[5]

In 2004, the American sociologist Nechama Tec provided a clear overview of her research into the motives of helpers. Just as Samuel Oliner, she was kept in hiding by a non-Jewish family in Poland during the war.

One of Tec's findings was that one of the shared characteristics in the group of helpers can be described as a strongly developed feeling of individualism. Such a characteristic (of which, most of these people are unaware) often gives individuals the ability to see beyond their immediate social surroundings.

In addition, the helpers were in many cases motivated by their own set of personal values. For this reason, many of the Jews they protected described them as courageous. The helpers, though, didn't think of themselves this way. Their often impulsive initiatives to help Jews were part of this mentality. Even when little or nothing was prepared in advance, they were still ready to help when a need arose.

Another characteristic of the helpers was that they acted from the feeling that they had to resist persecution and injustice. What others thought of that resistance was of no (or hardly any) interest to them. They followed an inner compulsion to help, as it were; this was the only thing that counted.

---

4 Michael Marrus, "The French Church and the Persecution of Jews in France," in *The Nazi Holocaust, part 8, volume 3*, ed. Michael Marrus (Westport: Meckler, 1998), 1284–1305. See also Moore, *Victims and survivors*, 199–201.
5 Eva Fogelman, *Conscience and Courage* (New York: Anchor Books, 1994), 1–393, mainly 161–220 and 58–61.

In this sense, the fact that it was Jews who were persecuted during the war was of minor importance. The helpers acted from a 'universal' perception. They did not see Jews, but simply defenseless and dependent people who needed help.[6]

The Non-Jewish Members of the Westerweel Group and their Motives

Despite the above-mentioned limitations of the research into the altruistic personality, it is enlightening to determine the extent to which the non-Jewish members of the Westerweel group satisfied the characteristics mentioned by Tec. In the case of Joop Westerweel, these characteristics seem almost to fit him like a glove. He was someone who was always ready to come to the aid of others. This was the case during his time in the Dutch East Indies, when he was confronted with the injustices of colonialism. Although his attempts to contact Indonesian nationalists were not successful, they were no less well intended. In his articles in the left-wing radical press, he sided with them completely.

With regard to Westerweel's time in Amsterdam, activities on behalf of others, both in his schoolwork and outside (such as for the korfball club *Blauw-Wit*), took a prominent place. The same can be said about his poorly paid work in Kees Boeke's school and his efforts in aid committees.

What others thought of his non-conformism did not appear to interest Westerweel very much. He told Pim van Maanen that the only reason he took part in the resistance was 'because I do not want to have to give something up in the war. If I want to sit on a terrace with a Jewish girl, then I want to be able to do that and nobody is going to stop me'.[7]

This opinionated willfulness also applied to his ideologically determined choice for 'free love', whereby he took no account of his wife's feelings. Other members of the Westerweel group also had sexual relationships with each other or lived together for periods of time. In this regard, the tensions of resistance work also played a part and, for the Palestine pioneers, naturally so did their age. Joop Westerweel's attitude with regard to 'free love' could, however, be called extreme.

---

[6] Nechama Tec, "Towards a Theory of Rescue," in *Making a Difference*, ed. David Scrase, et al., (Burlington, VT: Centre for Holocaust Studies University of Vermont, 2004), 22–48.
[7] GFHA, Cat. no. 143, Interview Yigal Benjamin-Willem van Maanen, 26 november 1989, 4.

At the end of 1930s, Westerweel started to help not only Jewish refugees, but also Spanish, Chinese and Finnish children who had been hard-pressed through the war. His 'universal' compulsion to help was certainly evident in May of 1940, when he took two heavily-wounded German parachutists to a medical first-aid station. During the German occupation, he gradually paid increasing attention to Jews, who were threatened by anti-Semitic measures.

What is noticeable here is that Westerweel initially operated mainly individually, apart from the less-successful pamphlet initiative at the end of 1940. Later Joop sent a protest letter to Seyss-Inquart, which had no results. He and Wil avoided cinemas, swimming pools and other similar places of leisure because Jews were not allowed there. He offered help to Jews in Amsterdam, but contacts with others from left-wing (radical) circles were apparently missing. It was only after what Fogelman has called the *enabling situation* (i.e., the requests for help, first from the Cohen family and shortly afterwards from Mirjam Waterman), that he was freed from his relative isolation and found a place in the resistance, which was gradually becoming organized.

There were several other group members who were involved in the 'hiding operation' in Loosdrecht from the very beginning: Bouke Koning, Jan Smit and Wil Westerweel. Of these, Bouke Koning was also almost a prototype of a nonconformist helper, as described by various researchers. He was a conscientious objector, he worked at the *Werkplaats* and he was ready to help Jews even before he was asked. Bouke was not only active in the Westerweel group, but also lent a hand to the Utrecht students' resistance. Despite the doubts he had about Westerweel's approach (in 1944, after his first arrest and detention), he did not hesitate to help him with taking pioneers across the border.

Bouke paid a heavy price for his resistance work. He returned from the camps as a wreck and never really recovered from the hardships that he had suffered.

Jan Smit more or less coincidentally joined forces with the Westerweel group, with whom he shared a left-wing background. As described, this background was formed in the Rotterdam Zuid community center, where he had been a member since the age of eight. Jan was of simple origin, an individualist who was not afraid to move outside his trusted milieu. He became increasingly annoyed by the German measures and was therefore happy that he was called upon to help the pioneers in Loosdrecht. Although he was sometimes sick with fear, he helped with all sorts of activities, such as the escapes from Westerbork.

Wil Westerweel differed very little from her husband with regard to character and philosophy of life. She was an idealist with a realistic view of people. She could possibly have started her resistance work earlier, just as her girlfriend Tine Segboer had done, if her family of three (later four) children had not been a hindrance. After she became involved with Joop in hiding the *Jeugdalijah* in Loosdrecht, she worked tirelessly in caring for them.

Just like Bouke Koning, Wil never got over her horrific experiences in the concentration camp. Her 'universal' attitude appeared in the 1970s and 1980s, evidenced by her critical attitude toward Israel's position regarding the Palestinian question.

Chiel Salomé was not involved in the Westerweel group from the beginning, but later became a very useful member. He and his brother were active in helping Jews before his contact with Joop Westerweel and the others. Chiel had begun this work on his own initiative, initially as a way to help a housemate and his relations. Later, Chiel continued because he thought that the persecution of Jews was unjust. A prison sentence of six weeks appeared to stimulate him to continue rather than to deter him in any way from this conviction. Chiel maintained the contacts with Sevenum for the Westerweel group, in which he had an auxiliary position.

Chiel Salomé also spent time in two concentration camps, where, with the necessary luck, he survived reasonably well. After the war, his altruistic personality was evident in the fact that, after a short recovery period, he went to work as supervisor for NSB youths whose parents (collaborators with the Germans) were in prison.

Tinus Schabbing was even more of an outsider than Chiel Salomé. Before he was active in the Westerweel group, he had accompantied a Jewish married couple to Belgium. He was a person who was upset by injustice (possibly motivated in part by the experiences of his difficult youth) and, for this reason, he took pioneers across the border. He hardly bothered himself with the rest of the organization.

Frans Gerritsen was certainly no outsider in the Westerweel group. However, he also had a separate position, which can probably best be described as the leader of the Haarlem section. Gerritsen played an important role in the group, but he did so mainly through the Palestine pioneers who were hiding with him and his wife—Max Windmüller, Lore Durlacher and later Paula Kaufmann. They acted as intermediaries with the Rotterdam section of the Westerweel group.

As far as is known, Frans Gerritsen had only one short meeting with Joop Westerweel. He seemed to prefer having as independent a position as possible—a preference that may have been motivated not only for reasons of safety, but also perhaps due to the fact that Joop Westerweel had a different, more passionate personality than he did. Frans Gerritsen likely realized that their differences in character could lead to frictions if they were to cooperate more closely. Gerritsen, the Haarlem pragmatic who was never arrested, operated more carefully than the Rotterdam section of the group, although it must be said that luck was also important when it came to safety.

Gerritsen had offered help to left-wing refugees and Jews before the formation of the Westerweel group. The Jews that he helped were initially acquaintances, but later he helped others as well. After starting his cooperation with the Westerweel group, he continued his work as a forger and hiding-place fabricator for other resistance organizations. In this he also had contact with the groups of the armed resistance. However, as a pacifist, he did not take part in their activities. Only once did he forego his principles to accept the help of some members of one of these groups in an attempt to derail a train near Westerbork.

Just like Joop Westerweel, Frans Gerritsen was a good example of an altruistic personality. Completely self-effacing, but with enormous courage, he gave himself fully in offering help to others, whoever they were.

And just as in various other resistance organizations, so-called half-Jews (a term devised by the Nazis) also played an important part in the Westerweel group. Gideon Drach, Norbert Klein, Kurt Reilinger and Ilse Birnbaum had this status and made good use of it. They didn't use their status to avoid persecution, as some Palestine pioneers in the same position did, but to be able to work more effectively for the Westerweel group.

This was a courageous choice, and the question is, were they aware of this? The four of them were active in the *Hechalutz* for years and appear never to have considered another possibility.

The Palestine Pioneers as Ideal Cooperation Partners

An important factor contributing to the relative success of the Westerweel group was the generally excellent personal relationship between the Jews and non-Jews in the group. Other important elements here were that most pioneers were middle class, and the older ones were often fairly well educated. In addition, many pioneers, as a result of their background in the Jewish youth movement, had things in common with the idiosyncratic pupils in the higher groups of the *Werkplaats*. For helpers such as Joop and Wil Westerweel, Jan Smit and Bouke Koning, and also for 'former *Werkplaats* workers' like Pim van Maanen, Gerda Loeff, Philip Rümke and Candia Boeke, this background will have provided them with a sort of recognition.

Moreover, the Westerweel group's shared socialist philosophy offered a good basis for cooperation. Van Maanen called socialism a 'common feeding ground'. 'Look, for us it meant mainly that we had to do something, that we could not simply let the situation go on. Something had to happen, and when these

Fig. 30: A group of Palestine pioneers in Barcelona in 1944. The group left Barcelona to go to Cadiz, from where they went to Palestine on the ship Guine. The little boy in the front row is Uri Durlacher.

boys arrived, we recognized something of ourselves in them, that we would be ashamed of if we were not successful.'

According to Van Maanen, he, and probably also other non-Jewish members of the Westerweel group, looked at the Palestine pioneers with a kind of admiration. The pioneers' 'enormous will to one's own free life' and their 'enormous will to survive' were highly impressive. This was certainly because many of the pioneers were often younger than the helpers and had already gone through so much. The non-Jewish members of the group regarded the Palestine pioneers as comrades with whom they shared a great deal.[8]

The non-Jewish members of the Westerweel group understood and respected the fact that the idealism of the pioneers was also aimed at the creation of a Jewish state in Palestine. But this ideal was further away from most of them than the socialist ideas that they shared.

---

8 Ibid., 6.

A Lack of Cooperation

There are probably two reasons why the Westerweel group, as an organization, had virtually no contacts with other resistance organizations. In the first place, when the group started in August 1942 there were almost no other organizations that the group could work with. The *Landelijke Onderduikcentrale* (LO, National Hiding Center) did not start to function meaningfully until the spring of 1943. Also, later, as far as is known, there were no contacts with the LO.

Mirjam Waterman approached the Utrecht students Olga Hudig and Hetty Voûte in the autumn of 1942. These two women would later play an important role in the Utrecht Kindercomité, which hid Jewish children. However, their efforts to find hiding places for the Loosdrecht pioneers did not produce any results.[9] Bouke Koning maintained contacts with students around this committee later in the war, but this was separate from his efforts for the Westerweel group.

One exception to this independent position appears to be Rotterdam, where a loose connection with an armed resistance group was realized from the middle of 1943, with Henk Brusse as contact. However, there was never any close cooperation. The members of the armed resistance group did lend a hand at various occasions and declared themselves ready to assist in the liberation of Joop Westerweel. However, because of the heavy guard on Joop, this came to nothing.

A second reason to avoid contacts with other resistance organizations was the betrayal of the flight to Switzerland of eight pioneers in the autumn of 1942. After this, the leaders of the Westerweel group decided to organize everything themselves, as much as possible. From the aspect of safety, this was sensible—and it also fit in with the non-conformist character of the group. However, contacts with other resistance organizations could also have provided valuable information.

For example, a group organized by the Amsterdam graphic designer Jaap Penraat used partly forged German documents from the end of 1942 or the beginning of 1943. Penraat had the documents certified by the *Prüfstelle* (inspection authority) on the Champs Elysees in Paris, almost a year earlier than the Westerweel group undertook any similar activity. According to his own statement, Penraat travelled 20 times with approximately 20 Jewish refugees to Lille. He thus handed over a total of about 400 refugees to the French resistance, who took them further to Spain. How Penraat found these contacts is, however, not

---

9 Bert-Jan Flim, *Omdat hun hart sprak. Geschiedenis van georganiseerde hulp aan Joodse kinderen in Nederland, 1942 – 1945* (Kampen: Kok, 1996), 53, 54.

clear.[10] His group operated very carefully and stopped travelling at the beginning of 1944 because of the danger of being betrayed.

Penraat's organization was small. However, the *Dutch-Paris* escape line built up by Jean Weidner, a Protestant Dutch textile merchant from Lyon, was huge. The organization, with escape routes to Vichy France, and (after the German occupation of Vichy in November 1942) to Switzerland and then Spain, had branches throughout the whole of France. But as far as is known, the Westerweel group never had any contacts with the *Dutch-Paris* group which, just as the Westerweel group, had Jewish and non-Jewish members. Weidner's group brought between 1000 and 1500 people, 800 of whom were Jews and more than 100 Allied pilots, to safety in Switzerland and Spain.

A comparison between the *Dutch-Paris* group and the Westerweel group cannot really be made, because of the differences in scale of operation and the social background of the members. Weidner's organization had dozens of Dutch, French and Belgian members and worked together with Dutch businessmen and diplomats who were staying in France. Later, the *Dutch-Paris* group was financed by the Dutch government in London, via Switzerland. The Westerweel group had fifteen core members and they, both non-Jews and Jews, came from different, more marginal sectors of society.[11]

In the autumn of 1942, Shushu Simon, who spoke with a heavy German accent, approached the Dutch vice-consul in Perpignan with a plan to smuggle Palestine pioneers to Libya via Spain. In Libya, they could join the British army. However, the vice-consul, who was possibly afraid of a German provocation, refused to cooperate. Still, the same diplomat, helped other Dutch refugees, both Jews and non-Jews.

Joop Westerweel preferred somewhat dubious Belgian smugglers like Rik Lelièvre and 'Theo' to take pioneers to France and Spain. Joop, the left-wing socialist who deeply distrusted everything that appeared to be 'authority', would probably not have wanted contact with businessmen and diplomats.

Kurt Reilinger helped to organize journeys to France several times, together with a students' organization from, most likely, Amsterdam or Utrecht. However, this was only to transport students to the building projects of the *Organisation Todt*. In France, the contacts were mostly quickly broken for reasons of safety.

---

10 Talbott, *Forging Freedom*. The Yad Vashem report about Penraat mentions only two transports. See also TV Program VPRO, Spoor Terug, Interview Jaap Penraat, dl. 1.
11 Megan Koreman, *Gewone helden. De Dutch Paris ontsnappingslijn 1942–1945* (Amsterdam: Boom, 2016). See also Hans Schippers, "Ontsnappen aan de nazi's," *Geschiedenis Magazine*, april 2018, 17–20.

The students mainly travelled with the transports to give the group a less 'Jewish' appearance.

During the French episode, the Westerweel group worked closely with the Jewish resistance (the *Armée Juive*, which was mainly comprised of East European Jews) and the groups around it. In France, these 'East-Jews' as they were called, often had the same separate social position as both the Jewish and non-Jewish members of the Westerweel group in the Netherlands. This cooperation proved successful in achieving the crossings to Spain. In Paris, however the contacts with the *Armée Juive* led to the betrayal of the Rue d'Erlanger, shortly before the liberation of the city, with serious consequences for members of both organisations.

Non-Conformist Resistance?

In conjunction with his altruistic attitude, Joop Westerweel could be called a pragmatic non-conformist. A non-conformist can be described as 'someone who strives for an independent attitude to life, and does not adapt'. Having an independent attitude to life indeed described Joop. However, when the goal demanded it, he adapted to the situation. From his position as a social outsider, and using his charismatic personality, he cooperated with various religious and social groups, as well as individuals—which was unusual in the socio-religiously segregated Dutch society of that time.

With the help of intermediary Chiel Salomé, Joop was able to gain the cooperation of the Catholic non-conformist Pastor Vullinghs and his helpers in finding dozens of hiding places for Jews in North Limburg. In Friesland, and also elsewhere, Joop Westerweel worked closely with Protestant ministers to hide Palestine pioneers. Other pioneers were housed with socialists and communists. In his cooperation with Protestant Christians, he had the advantage that he was familiar with Protestant idiom and customs, due to his background with the Plymouth Brethren. However, this same background also ensured that he was an outsider with regard to the established church communities.

The description of the non-conformist (but with positive connotations) also fits well with the aforementioned characteristics of the altruistic personality, as described in Nechama Tec's overview. A strongly developed feeling of individualism often placed the people involved outside their social surroundings, with their own set of personal values, and unaffected by the opinion of others; there appears to be hardly any difference between this personality and positively-oriented non-conformism.

Several members of the Westerweel group, certainly the non-Jewish members, but also some of the older Jewish Palestine pioneers (who were influenced by the 'new style' of the youth movement), can be characterized as left-wing non-conformists. As such, it is useful to categorize the group as broadly non-conformist.

Non-Conformism and Lack of Professionalism

The next question to ask is whether this non-conformism also influenced the Westerweel group's resistance activities. But giving a straightforward answer to this question is not possible. Resistance is resistance, and certain rules of conduct must be observed in order to be able to operate effectively and safely.

With regard to normal practice in the resistance movement, these non-conforming characteristics therefore did not separate the Westerweel group from other resistance organizations. However, a difference must be noted between the Jewish and non-Jewish members of the group. The non-Jewish members ran less risk and could move about more freely. Joop Westerweel repeatedly ignored the usual rules of conduct that were based on caution. An example is the train journey he made at the beginning of 1944 with Lore Sieskind to Toulouse. During this journey, Joop, who had chosen to travel in a German army train compartment with false papers, spent the entire night talking to several soldiers. He tried to convince them of his pacifistic and anti-fascist views. To Lore's great relief, they were not arrested or thrown out of the train. Upon reflection, she even thought that his words might have had a positive effect.[12]

Joop was also the inspiring figure who, at the right times and perhaps sometimes at the wrong times, read poems out loud or quoted from the Bible. 'He could not resist the temptation; he had to convince you that this is how we must live. He could drag you along and you did not have the heart to say 'no'.' According to Willem van Maanen, these were preachers' traits, which he likely had inherited from his father.[13]

Several group members, annoyed by Westerweel's well-intentioned but dangerous attempts to convince people of his views, refused to travel with him. Most of them operated more carefully—in any case trying to avoid attention.

Despite the best efforts of several individuals in the group, the group's activities generally continued to lack professionalism. In the spring of 1944, crossing

---

12 Lore Zimmels (Sieskind), "Lore Zimmels," Westerweel, *Verzet*, 98.
13 GFHA, Cat. no. 143, Interview Benjamin-Van Maanen, 5.

the border to Belgium was carried out in an even riskier manner than the previous year. Joop Westerweel continued to use the dangerous route through Budel, although he and others were often arrested there. He had to talk his way out of it each time, with all the associated risks, which he had been warned about beforehand. Also, near a border crossing south of Breda, Menachem Pinkhof at one time lost contact with two pioneers, who were then arrested by a German patrol as a result. The Westerweel group had apparently only limited relations, if any, with local resistance groups in the border villages

This rather poor professional character was also evident in the fact that the group's members were not sufficiently cautious about possible betrayal and infiltrations. The damage caused by Mr. and Mrs. De Ruiter, for example, could have been prevented by being more cautious with offers of help. The group was also warned several times about Karel Kaufmann, but the Rotterdam section was apparently not of aware of this, which set off all of the consequences surrounding the trap set up around Lettie Rudelsheim.

The infiltration of 'Karel' in the spring of 1944 revealed several mistakes. Without knowing him, but based on a note from Joop Westerweel, Willem and Gerda van Maanen took him into their house. 'Karel' was present at meetings planning Joop's liberation and thus became aware of the plans and the people who were involved. Van Maanen did move when he became suspicious, but did not warn the others. Jan Smit and Lore Durlacher escaped arrest only by coincidence.

Another example of the continuing lack of professional character after almost two years of activity was a meeting organized at a camping site in Nunspeet at the beginning of April 1944. Jan Smit, Menachem Pinkhof, Mirjam Waterman and Tinus Schabbing and, from France, Paula Kaufmann, Kurt Reilinger and Max Windmüller, among others, were present.

The primary objective of the meeting was to discuss the group's position after the arrest of Joop Westerweel and Bouke Koning. However, relaxation and team building were also on the programme. Pancakes were made, games were played and the hora (a Jewish folk dance) was practiced. Tinus Schabbing 'really liked' the multi-day camping party, and photographs were also taken. Later, he realized that it had been 'a bit dangerous'. A police raid would have meant the end of the entire Westerweel group, including the French section.[14]

Non-conformism appeared to be a clear disadvantage when operating abroad. For example, no contact was sought with other organizations, such as the *Dutch-Paris* group, which was active in Belgium and France and also smug-

---

14 Interview Sytske De Jong-Tinus Schabbing, 24.

gled refugees to Spain. The reason behind this was a striving for safety and a certain degree of social isolation, plus also a desire for independence which would allow them to build their own escape routes and cooperate with the *Armée Juive*.

The non-conformism of the Jewish group's members almost gave the impression that they were thumbing their noses at the enemy, in a manner similar to literary hero Reynard the Fox. To what extent they were influenced in this by Joop Westerweel is not certain, but it is probable that his influence was significant. The harmonica tunes that Max Windmüller played on train journeys between Brussels, Paris and other locations in France were mainly intended to distract the attention and calm down the often nervous pioneers whom he accompanied. But the 'Palestinian' folk songs that he wove through his repertoire were a daring provocation.

One could say the same about visits to the Paris Opera, where pioneers, dressed in civilian clothes, sat down in pairs of two between uniformed Germans. Several of the Palestine pioneers living in France often used false German papers to travel in German army wagons, to eat in German canteens, to visit German army cinemas and to go on leave to the Netherlands. Youthful overconfidence, a feeling of invulnerability and an underestimation of the opponent were partly responsible for this behavior. However, the pioneers also wanted, perhaps unconsciously, to debunk the prejudice of the cowardly Jew and to show that they were not afraid of their anti-Semitic persecutors.[15]

Non-Conformism, Pacifism, and Emancipation

Non-conformism in the Westerweel group also played a part in two other areas: pacifism and the position of women.

For people like Joop and Wil Westerweel, Bouke Koning and Frans Gerritsen, pacifism was a matter of principle. They completely rejected the use of weapons. Westerweel also had significant ideological problems with the work for the *Organisation Todt* that many pioneers in France had to do of necessity. He urged Kurt Reilinger to find alternatives and supported pioneers who did not want to work on the *Atlantic Wall*.

Other members of the group were less rigid. In the initial period of hiding people, the Westerweel group had accepted the care of an older teacher. The teacher's behavior created a danger for his helpers, and twice the group discussed whether he should be liquidated.

---

[15] Latour, *Résistance*, 108. See also Asscher, *Van Mokum*, 180–181. Siegel, *Locomotieven*, 151.

The first time, resistance man Jaap Lambeck from Loosdrecht spent hours digging in his orchard 'to get a hole big enough for a body'. However, the execution was cancelled at the last moment. Lambeck was told that the teacher had been put on an Argentinian ship, 'his destination South America'.[16] The second time, Jan Smit and Norbert Klein decided not to go through with a liquidation plan using a Rotterdam armed resistance group. These liquidation plans were far from pacifistic in nature. It is not known if Joop Westerweel or Bouke Koning knew of them.

Other members of the Westerweel group also diverted from the pacifistic principles of Westerweel and those around him. Kurt Reilinger kept a firearm in the Netherlands, which was probably the same gun that he had in Paris. During the crossing to Spain, not only the guide Adrien, but also pioneers such as Paul Siegel were armed. As described, several other pioneers in France were involved in the *Maquis* or the *Armée Juive* and later in the armed resistance. One could say that the use of weapons was unavoidable in certain situations. However, this does not affect the principal attitude of the pacifistic part of the group.

The position of several women in the Westerweel group shows a clear break with the established gender roles of men and women. Lore Durlacher, Wil Westerweel, Lore Sieskind and Mirjam Waterman had positions in the organization that were equivalent to those of the men. The same also applied later in France to Paula Kaufmann and Lolly Eckhardt, among others.

The noticeable thing here is that, in standard situations, the Westerweel group's members reverted to the traditional man-woman distribution of labor. In the safe houses, for example, women did the housekeeping. Lore Sieskind, 'who had stood her ground' in the Dutch resistance, worked in the kitchen in the *Beth Chaloets* in northern France.

In cases where the usual man-woman pattern was broken, the Westerweel group showed similarities with the communist resistance. Here also, women fulfilled tasks that at that time could only be described as 'masculine'. This became usual, however, only after the women fought for equality. Hanny Schaft and the Oversteegen sisters were even involved in the liquidation of collaborators.

The Westerweel group was forced to all but end its activities in the summer of 1944, just when other forms of resistance to the Nazis were increasing in size and strength. The persecuted Jews, however, were not able to profit from this increase in resistance, since the Nazis had already transported the great majority of them to Poland, where they had been murdered. The Palestine pioneers who

---

16 NIOD, Jaap Lambeck, Herinneringen van een voormalig verzetsleider, Typescript, zonder plaats, zonder jaar, 49.

were helped by the Westerweel group stayed mostly in France and Spain. A smaller number remained in hiding in the Netherlands and was looked after mainly by the local resistance.

During 1943, the Westerweel group was taking care of about 300 Palestine pioneers in hiding. No one can say exactly how many of them survived the war. In the Netherlands and in France, the group was joined by other pioneers who had first gone into hiding or had fled individually. Several pioneers were arrested in both countries. In the Netherlands, these were mainly arrests at border crossings to Belgium, around the Rotterdam safehouse in October 1943 and in The Hague in April 1944. Of the approximately twenty arrested pioneers, about half survived deportation to the east.

In France, most of the arrests occurred in Paris during April and August 1944. Other than might be expected, however, most of the—again, approximately twenty—detainees survived the German concentration camps. It is certain that one Palestine pioneer died in an accident in the crossings to Spain.

With some caution, it can be said that the Westerweel group helped between 250 and 275 Palestine pioneers to escape the Nazi persecution. In addition to this, several dozen of other Jews were taken to hiding places and put into the care of the local Dutch resistance. The members of the Westerweel group, Jews and non-Jews, might be called heroes, but they certainly would have rejected this name. Together they had travelled 'the right path', to quote Joop Westerweel,' and that was enough.

Bibliography

Archives

Nederlands Instituut voor Oorlogs- Holocaust- en Genocide Studies (NIOD), Amsterdam
    Collectie, Westerweelgroep, Doc. II – 249–0296 A
    Collectie, Palestina Pioniers, Doc II – 614 A (A+B)
    KB I – 4049 – Koning, Bouke
    KB I – 10684 – Maanen, Willem G.
    KB I – 7385 – Westerweel, Joop

Ghetto Fighters House Archives, (GFHA), Lohamei Hageta'ot, Israel
    Holland Section
    Photo Archives

Studie- en Documentatiecentrum Oorlog en Hedendaagse Maatschappij (Cege-Soma), Brussel, België
    File H.A.M.G. Lelièvre

Nationaal Archief, Den Haag
    Archief van het Ministerie van Koloniën
    Dossier J.H. Westerweel
    Centraal Archief Bijzondere Rechtspleging (CABR)
    Inventarisnummers: 93279, 73075, 45502
    NBI Inventarisnummer: 5025

Mémorial de la Shoah, Centre de documentation juive contemporaine, Paris
    Collection Armée Juive

Yad Vashem, Jerusalem, Israel
    Righteous Database
    Members and helpers of the Westerweel group, Jaap Penraat

Stadsarchief Rotterdam
    Archief gemeentepolitie Rotterdam 1944, no. 63
    63/3781 Formulieren ter opname van gearresteerde personen in bewaring, 1944 Hop – Ka
    Arrestanten kaarten no. 8 A – Jong t/m Klaassen nrs. 1374–1377

Private Collections
    B. Westerweel, Zutphen
    Ph. Rümke, Amsterdam

Verbal and email information
    G. Smit
    B. Westerweel
    Ph. Rümke
    M. Bouman-Gerritsen
    Informatiecentrum Kamp Westerbork
    Rode Kruis Oorlogsarchieven, Den Haag

Literature

Abuys, Guido, en Dirk Mulder. *Een gat in het prikkeldraad*. Hooghalen/Assen: Herinneringscentrum Kamp Westerbork, 2003.
Alpert, Michael. "Spain and the Jews in the Second World War." *Jewish Historical Review* 42 (2009): 201–210.
Asscher, Berrie. *Van Mokum naar Jeruzalem (1924–1944)*. Beersheba: self-published, 1996.
Asscher, Channa. "Channa Asscher." In *Verzet zonder geweld: Joop Westerweel 1899–1944*, edited by Willie Westerweel. self-published, 1964. 88, 89.
Asscher, Naftali (Harry). "Naftali Asscher" In *Verzet zonder geweld: Joop Westerweel 1899–1944*, "Ibid.". 87–89.
Avni, Haim. "The Zionist underground in Holland and France and the escape to Spain." In *Rescue attempts during the Holocaust*, edited by Yisrael Gutman and Efraim Zuroff. Jerusalem: Yad Vashem, 1974. 555–590.
Baron, Lawrence. "Dynamics of Decency." In *The Nazi Holocaust*, edited by Michael Marrus. Munich: Piper, 1989. Part 5, vol. 2, 608–628.
Ben Heled-Rudelsheim, Letty, en Miriam Dubi-Gazan. *Gesprekken met mijzelf in Auschwitz*. Kampen: Kok, 2003.
Benjamin, Yigal. *They were our friends. A memorial for the members of the Hachsharot and the Hechalutz underground in Holland murdered in the Holocaust*. Jerusalem: Association of former members of the Hachshara and Hechalutz underground in Holland, 1990.
Bergier, Jean-François, ed. *Switzerland, National Socialism and the Second World War, Final Report*. Zurich: Pendo-Verlag, 2002.
Boeke-Cadbury, Betty. "Joop en de Werkplaats" In *Verzet zonder geweld: Joop Westerweel 1899–1944*, "Ibid.". 15–18.
Boeke, Julia, en Cees Smit. *Archief Werkplaats Kindergemeenschap (Bilthoven) Inleiding Geschiedenis*. Amsterdam: International Institute of Social History, 1990. 1–9.
Boom, Bart van der. *"Wij weten niets van hun lot." Gewone Nederlanders en de Holocaust*. Amsterdam: Boom, 2012.
Bouter, Muis de. "Ik sta bij hem in de schuld" In *Verzet zonder geweld: Joop Westerweel 1899–1944*, "Ibid.". 31.
Braber, Ben. *This cannot happen here: integration and Jewish resistance in the Netherlands, 1940–1945*. Amsterdam: Amsterdam University Press, 2013.
Brouwers, Wilko. "Ward in Nederland, dan en nu." Accessed June 3, 2013. https://www.wardcentrumnederland.eu.
Browning, Christopher. *The Origins of the Final Solution. The Evolution of Nazi Jewish Policy, September 1939-March 1942*. Lincoln: Nebraska University Press, 2004.
Bruin, André de. *Noord Limburg integraal bekeken, 1850–1950: zoektocht naar de wortels van een cultuur*. Sittard: Mooi Limburgs Boekenfonds, 2010.
Bunge, Gary M. "Christian Zionism, Evangelicals and Israel." Accessed February 22, 2013. https:// www.christianzionism.org.
Cammaert, A.P.M. *Het verborgen front. Geschiedenis van de georganiseerde illegaliteit in de provincie Limburg tijdens de Tweede Wereldoorlog*. Leeuwarden: Eisma, 1994. Deel 1.
Canter Visscher-Kolff, Greet. "Herinneringen van een Montessorileidster." In *Verzet zonder geweld: Joop Westerweel 1899–1944*, "Ibid.". 61, 62.
Chaigneau, J-F. *Le dernier wagon*. Paris: Julliard,1981.

Dam, M.J. van, H.C. van Iterzon, en H.J. Maarsingh. *Gouda in de Tweede Wereldoorlog.* Delft: Eburon, 1995. 169–184.
Desquesnes, Rémy. "L'Organisation Todt en France (1940–1944)." *Histoire, économie et société* 3 (1992): 535–550.
Dijk, Cor van. "Joop Westerweel." Blauw-Wit 50 (jubilee volume) (January 1967): 7–8.
Dunk, H.W. von der. *Terugblik bij strijklicht. Jeugdherinneringen.* Amsterdam: Bakker, 2008.
Dunk, Hermann von der. "Hij was overal." In *Verzet zonder geweld: Joop Westerweel 1899–1944,* "Ibid.". 19–22.
Flim, Bert-Jan. *Omdat hun hart sprak. Geschiedenis van de georganiseerde hulp aan Joodse kinderen in Nederland, 1942–1945.* Kampen: Kok, 1996.
Flörsheim, Hans. *Über die Pyrenäen in die Freiheit.* Konstanz: Hartung-Gorre Verlag, 2008.
Fogelman, Eva. *Conscience and Courage: Rescuers of Jews during the Holocaust.* New York: Anchor Books, 1994.
Frank, Anne. *Het Achterhuis.* 16nd ed. Amsterdam: Uitgeverij Contact, 1957.
Gasenbeek, Bert, en Chris Hietland. *Van jeugdig pacifisme naar geestelijke weerbaarheid. De Jongeren Vredes Actie (1924–1940).* Breda: Papieren Tijger, 2012.
Giebels, Ludy. *De Zionistische beweging in Nederland 1899–1941.* Assen: Van Gorcum, 1975.
Gitelman, Zvi. *A Century of Ambivalence. The Jews of Russia and the Soviet Union, 1881 to the Present.* New York: YIVO Institute for Jewish Research, 1988.
Goren, Ora (Durlacher, Lore). "Ora Goren." In *Verzet zonder geweld: Joop Westerweel 1899–1944,* "Ibid.". 101–102.
Hetkamp, Jutta. "Die jüdische Jugendbewegung in Deutschland von 1913–1933." PhD. diss., Universität Essen, 1994.
Holten, F. van. "Een eenzame man in de kerk." In *Reformatorisch Dagblad,* 7 december 2000, p. 17.
Hooghiemstra, Daniela. "De geest in dit huis is liefderijk. Het leven en de Werkplaats van Kees Boeke (1884–1966)." PhD. diss., Universteit van Amsterdam, 2013.
Jong Fr. de. "Kees Boeke." In *Onderwijskundigen van de 20ste eeuw,* eds Q. van der Meer en H. Bergman. Amsterdam/Groningen: Intermediair, 1979. 142–158.
Jong, L. de. *Het Koninkrijk der Nederlanden in de Tweede Wereldoorlog.* Deel 5, eerste helft, tweede helft. Den Haag: Martinus Nijhoff, 1974.
Jong, L. de. *Het Koninkrijk der Nederlanden in de Tweede Wereldoorlog.* Deel 6, eerste helft, tweede helft. Den Haag: Martinus Nijhoff, 1975.
Jong, L. de. *Het Koninkrijk der Nederlanden in de Tweede Wereldoorlog.* Deel 7, eerste helft, tweede helft. Den Haag: Martinus Nijhoff, 1976.
Jong, L. de. *Het Koninkrijk der Nederlanden in de Tweede Wereldoorlog.* Deel 11a, tweede helft. Den Haag: Martinus Nijhoff, 1984.
Kooistra, Jack, en Albert Oosthoek. *Recht op wraak. Liquidaties in Nederland 1940–1945.* Leeuwarden: PENN Uitgeverij, 2009.
Koreman, Megan. *Gewone helden. De Dutch-Paris ontsnappingslijn 1942–1945.* Amsterdam: Boom 2016.
Laqueur, Walter. *A History of Zionism.* London: Weidenfeld & Nicolson, 1976.
Laqueur, Walter. *Young Germany: A History of the German Youth Movement.* New Brunswick/London: Transaction Books,1984.
Latour, Anny. *La Résistance juive en France (1940–1944).* Paris: Stock, 1970.
Lazare, Lucien. *La résistance juive en France.* Paris: Stock, 1987.

Leuvenberg-Nathans, Jochebed. *Twee Palestina-pioniers in oorlogstijd.* Winsum: Profiel, 2000.
Liempt, Ad van, en Jan Kompagnie, *Jodenjacht. De onthutsende rol van de Nederlandse politie in de Tweede Wereldoorlog.* Amsterdam: Uitgeverij Balans, 2011.
Lind, Jakov. *Stap for stap. Autobiografie.* Amsterdam: De Bezige Bij, 1970.
Longerich, Peter. *Politik der Vernichtung: Eine Gesamtdarstellung der nationalsozialistischen Judenverfolgung.* Munich: Piper, 1998.
Luiten, Hans. "De clubgeschiedenis van Blauw-Wit 1916–1941." BA Thesis, Universiteit van Amsterdam, 2000.
Maanen, Willem van."Verzet zonder geweld." *Maatstaf* 12 (1964): 2–8.
Mageen, N. *Van zonsondergang tot dageraad.* Beersheba: self-published, 2002.
Marrus, Michael. "French Church and the Persecution of Jews in France." In *The Nazi Holocaust, historical articles on the destruction of Europian Jews,* edited by Michael Marrus. Westport: Meckler, 1989. Part 8, vol. 3. 1284–1305.
Mechanicus, Philip. *In Dépôt, dagboek uit Westerbork.* Amsterdam: Van Gennep, 1964.
Meyer-Dettum, Klaus. *Max Windmüller 1920–1945.* Emden: Arbeitskreis Juden in Emden, 1997.
Moore, Bob. *Victims and survivors. The Nazi persecution of the Jews in the Netherlands.* London: Arnold, 1997.
Moore, Bob. *Survivors. Jewish Self-Help and Rescue in Nazi-Occupied Western Europe.* Oxford: University Press, 2010.
Noordergraaf, Herman. "Ligt, Bartholomeus de" In *Biografisch Woordenboek van het Socialisme en de Werkersbeweging.* Amsterdam: International Institute of Social History, 1988. Deel 3. 123–126.
Noordergraaf, Herman. "Bergh van Eysinga, H.W.P.E. van den."In *Biografisch Woordenboek van het Socialisme en de Werkersbeweging.* Amsterdam: International Institute of Social History, 1988. Deel 3. 10–12.
Oliner, Samuel and Pearl. *The Altruistic Personality: Rescuers of Jews in Nazi Europe.* New York: Free Press, 1988.
Ouweneel, Willem J. *De Vergadering van Gelovigen.* Kampen: Kok, 2002.
Payne, Stanley. *Franco and Hitler: Spain, Germany, and World War II.* New Haven/London: Yale University Press, 2008.
Pinkhof, Menachem. "Kurt Reilinger." *Dawar Hechalutz* 1 (oktober 1945): 1.
Pinkhof, Mirjam, en Ineke Brasz, eds. *De Jeugdalijah van de Paviljoen Loosdrechtsche Rade, 1939–1945.* Hilversum: Verloren, 1998.
Plantinga, Sierk, "Joseph William Kolkman (1896–1944) en de Engelandvaarders." In *Negende Jaarboek van het Rijks Instituut voor Oorlogsdocumentatie.* Zutphen: Walburg Pers, 1998. 10–36.
Poznanski, Renée. *Jews in France during World War II.* Hanover, NH: University Press of New England, 2001.
Presser, J. *Ondergang. De vervolging en verdelging van het Nederlandse Jodendom,* Den Haag: Staatsuitgeverij, 1965. Deel 1, 2.
Rens, Herman van. *Vervolgd in Limburg. Joden en Sinti in Nederlands-Limburg tijdens de Tweede Wereldoorlog.* Hilversum: Verloren, 2013.
Samson, Schlomo. *Zwischen Finsternis und Licht. 50 Jahre nach Bergen-Belsen.* Jerusalem: Rubin Mass GmbH, 1995.
Schippers, Hans. Ontsnappen aan de nazi's. In *Geschiedenis Magazine,* (april 2018): 17–20.

Siegel, Paul. *Locomotieven trekken wagons 1933–1945*. Westervoort: Van Gruting, 2000.
Stam, Jaap. "Maatschappelijke vorming door gemengd sporten." *Volkskrant*, 8 november 2010, p. 6.
Stegeman, H.B.J., J.P. Vorsteveld met een bijdrage van J.W. Reutlinger, *Het Joodse Werkdorp in de Wieringermeer, 1934–1941*. Zutphen: Walburg Pers, 1983.
Stern, Richard. "Joop Westerweel." *Levend Joods Geloof*, (mei 1984): 13.
Storm, Bram. "Droom en daad." In *Verzet zonder geweld: Joop Westerweel 1899–1944*, "Ibid.". 35, 36.
Straaten, Frans van der. *Om nooit te vergeten. Palestina-pioniers in Nederland 1939–1945*. Mijnsheerenland: self-published, 1998.
Talbott, Hudson. *Forging Freedom*. New York: G.J. Putnam's Sons, 2000.
Tec, Nechama. "Towards a Theory of Rescue." In *Making a Difference. Rescue and Assistance During the Holocaust*, edited by David Scrase. Burlington, VT: Centre for Holocaust Studies University of Vermont, 2004. 22–48.
Wasserstein, Bernard. *Gertrude van Tijn en het lot van de Nederlandse Joden*. Amsterdam: Nieuw Amsterdam, 2013.
Westerweel, J.G. "Het doekje voor het bloeden." *Bevrijding* 60 (september 1925): 193, 195.
Westerweel, J.G. "Pasifisme." *Bevrijding* 63 (december 1925): 219.
Westerweel, J.G. "Onze muziek speelt door... dans op haar maat." *Bevrijding* 65 (februari 1926): 235, 236.
Westerweel, J.G. "Tramstaking Zutfen – Emmerik." *Bevrijding* 66 (maart 1926): 240.
Wyman, Mark. *DPs: Europe's Displaced Persons, 1945–1951*. Philadelphia: Balch Institute Press, 1989.
Zimmels (Sieskind), Lore. "Lore Zimmels." In *Verzet zonder geweld: Joop Westerweel 1899–1944*, "Ibid.". 95–99.
Zuylen, L.F. van. *Palestinapioniers in Twente 1933–1945. Een vergeten hoofdstuk*. Enschede: Twente Akademie, 1995.

Unpublished sources

Jong, Sytske de. "De Westerweelgroep." MS thesis Rijksuniversiteit Groningen, 2001.
Westerweel, Wil. *Lijn of cirkel*.
Stamberger, Janiv. "Be Strong and Brave! A small youth movement in a sea of history. The Hashomer Hatzair Antwerp (1920–1948)." MS thesis University Gent, 2013.

TV recording

Spoor Terug, Interview with Jaap Penraat Deel 1, VPRO, 17 juni 2001.

Glossary of Terms

Agudah	Strictly orthodox movement with a kibbutz in Enschede.
Aliyah	Jewish emigration to Palestine.
Armée Juive	Jewish resistance organization in France. Founded in 1942 by Avraham Polonski and Aron Lublin. From mid-1944, the Armée Juive called itself Organisation Juive de Combat.
Betar	Nationalistic Zionist youth organization linked to the Revisionistic movement of Ze'ev Jabotinski.
Beth Chaloets	Shelter for Palestine pioneers.
Chalutzim	Ivrit word for Palestine pioneers.
Chaverim	Ivrit word for mates, comrades.
Habonim	Socialist-orientated Zionist youth organization.
Hachshara	Preparation period and practical training in mainly agriculture and horticulture for settling in Palestine.
Hashomer Hatzair	Left-wing socialist Zionist youth organization.
Hechalutz	Umbrella organization of Palestine pioneers.
Jeugdalijah	An organization whose purpose was to train Jews up to and including 18 years of age for settling in Palestine.
Jewish Brigade	British army unit founded in 1944, consisting mainly of Palestinian Jews. Disbanded in 1946.
Joint	American-Jewish aid organization. The complete name is: American Jewish Joint Distribution Committee.
Joodse centrale voor Beroepsopleiding (JCB)	Organization that co-ordinated the different Jewish vocational training courses, realised In June 1940.
Joodse Jeugd Federatie (JJF)	Umbrella organization of Jewish youth movements founded at the beginning of the 1920s.
Kibbutz, plural Kibbutzim	Collective farm.
Landwacht	A paramilitary auxiliary service of the German occupier, founded the end of 1943, mainly comprising members of the NSB (Dutch Nazi party).
'Les Hollandais'	Palestine pioneers from the Netherlands staying in France.
Maquis	Guerrilla units of the French resistance.
Maquisards	Members of the Maquis
Mizrachi	Moderate religious Zionist organization that had *hachshara* kibbutzim in Beverwijk, Franeker and later Laag-Keppel.

Palestine pioneers	Young Jews who followed an (agriculture) course from shortly after the First World War to be well prepared to go Palestine.
Poale Zion	Socialist Zionist organization.
Torah	The first five books of the Old Testament.

Index

Aafjes, Bertus 74
Adrien, mountain guide 219, 222, 225f., 263
Ahlfeld, Werner 90, 102f., 168, 172, 174, 183
Akker, family van den 156
Andriesse, Joop 229, 237
Appenzeller, Ernest 232–234
Aschheim, Bernhard 82
Aschheim, Dov 29
Asscher, Abraham 25
Asscher, Berrie 105, 107, 142, 144, 154, 156f., 193, 221, 223, 225, 227
Asscher, Ernst 156, 232f., 235
Asscher, Hanna 239
Asscher, Harry 93, 115f., 138f., 188, 192, 198, 213, 217, 239
Avni, Haim 123

Bachrach, Fritz 127
Bandmann, Hans 93
Baron, Lawrence 250f.
Bendiks 134
Benjamin, Emanuel 155
Benjamin, Kurt 89, 124, 126, 128, 133, 227, 229
Benjamin, Walter 125
Benjamin, Yigal 3, 18, 242f.
Berg van Eijsinga, Henri van den 35, 41
Birnbaum, Ilse 102, 113, 116, 118, 255
Bloch, Marcel 234
Blüth, Erika 21f., 29f, 71, 170
Blüth, Kurth 30, 172, 208
Boeke, Candia 67, 80, 79, 255
Boeke, Helen 56
Boeke, Kees 32, 39, 45–49, 53–57, 59, 61, 65, 176, 252
Boeke-Cadbury, Betty 32, 39, 45, 46
Bonn, Max 154f.
Bonn, Rie 154, 155
Borochov, Ber 8, 10
Bosdriesz, Wil (Willie) 42f.
Bouter, Muis de 49f.

Boutet, Eugénie 135, 137, 139, 157
Brasz, Ineke 243
Brekveld, boer 111
Britz, Betty 20, 217, 230
Bruin, Linnie de 90, 102f.
Brunner, Alois 234
Brusse, Henk 203, 206, 210, 257

Cammaert, A.P.M. 138
Campers, H. 83
Campert, J. 212
Coevorden, Sophie (Adina) van 77, 88, 120, 218
Cohen, David 15, 16, 25
Cohen, family 253
Cohen, Lodi 20, 30
Cohen, Rachel 117, 190
Cohen, Ru (Rudolf) 16, 18, 19, 25, 90, 112, 117, 148, 155, 164, 190
Cole, Julia 132f., 143
Cole, Maria 132
Cosmann, Ernst 102, 103
Craenenbroek, van (company) 156

Daams, Jasper 119, 131, 177
Darby, John Nelson 34
Davids, Aaron 15
Decker, Helga 187, 242
Deppner, Erich 211
Deutsch, family 19
Direktor, Ruth 200f., 209
Domela Nieuwenhuis, F. 67
Drach, Gideon (Thomas) 88, 92–96, 98, 100, 109, 141, 179, 180, 182, 187f., 255
Dubowski, Alfred 106
Dunk, Hermann von der 4, 47, 50f., 51, 55f.
Durlacher, Heinz 213, 226
Durlacher, Lore 109f., 114f., 146f., 147, 171, 173f., 176–178, 181, 182, 188, 197, 206, 209, 212, 217, 241, 254, 261, 263
Durlacher, Ruth 226

Durlacher, Uri 226
Duyn, J.F. van 165

Eckhardt, Lolly 217, 230, 234, 241, 263
Edelstein, Jakub 25–27
Ehrlich, Hans 150, 151, 217, 230, 232–234
Ehrlich, Kurt 126, 128, 150, 227
Eichmann, 234
Einstein, A. 41
Enckevort, Jan and Nelly van 138, 140
Engel, Schraga 29, 145

Fischer, Emile 16
Fischer, Franz 195
Fischer, Schlomo 16
Flörsheim, Hans 102, 107–110, 155, 220, 222, 226
Fogelman, Eva 251, 253
Franco, General 120, 125, 226
Fränkel, Alfred/Zippi 168f., 174, 217, 224f., 230, 232–234, 236, 240f.
Fränkel, Heinz 150–152, 154, 158, 208
Freier, Recha 19

Gandhi, M. 41
Gemmeker, Albert 159, 163, 170f., 208
Gerler, Willy 127f.
Gerritsen, Frans 96–101, 115f., 147f., 163, 171, 173, 176–179, 181f., 187f., 197, 199, 203, 212f., 217, 241, 242, 254f., 262
Gerritsen, Henny 97, 100, 212, 241f.
Giniewski, Otto 122
Glèbe d'Eu, Guy (Captain Jack) 232
Glücker, Emil 106, 142, 152, 154, 166, 221
Gordon, A.D. 10

Hannemann, Kurt 87–90, 92–96, 98, 100, 102, 105, 108f., 140, 164f., 186f., 190, 194, 197
Harinxma thoe Slooten, A. van 54
Heeckeren, Willem van 131, 201
Heiman, Mrs. 109f.
Heinrich, Arthur 28, 76, 142, 153
Heinrich, Joseph 29
Hermann, Susi 230, 234
Herz, Heinz 114f., 141

Herzberg, Abel 15, 117, 209
Herzl, Theodor 6
Hess, Fiet 229
Hirsch, Ernst (Willy) 142f., 147–149, 153–156, 158, 211, 217, 222f., 230f., 233, 235
Hitler, Adolf 19, 26f., 51, 111, 147, 153, 211
Hoffman, SD 196
Homme, Co 212
Hooghiemstra, Daniela 56
Horowitz, Hans 184
Hudig, Olga 257

Italiaander, Herman 161, 167, 172–174, 177

Jabotinsky, Vladimir (Ze'ev) 10, 218, 221
Jakobs, Ludwig 232, 234
Jarblum, Marc 122, 218
Jefroykin, Jules (Dika) 231
Jesse, Bob 96f., 100, 212
Jesse, Dientje 97
Jong, A.R. de 39
Jong, L. de 247

Kahn, Werner 142
Kann, Jacobus 15
Kapel, René 232–234
'Karel' 205–208, 212, 214
Kaufmann, Karel 179, 191, 193–195, 202, 204f., 261
Kaufmann, Paula 100, 141, 156, 157, 217, 232–235, 254, 261, 263
Kettner, Lily 22, 138f., 170f., 238–240
Klein, Norbert 77, 92, 93, 100, 108, 109, 127, 141, 157, 186f., 190, 212, 255, 263
Kolff, Greet 197
Kolkman, Joop 122
Koller, Arnold 184
Koning, Bouke 30, 66f., 69, 72, 79, 83, 146, 188, 190, 199f., 208, 210, 253–255, 257, 261–263
Kraan, Hendrika (Riek) L.W. 35, 42
Kramer, Froukje 67
Kroon, Dirk 197
Krzeszower, Max 146

Lambeck, Jaap 67, 70, 75, 263

Landauer, Paul 155
Lande, Metta 76, 217, 230, 232f., 241
Lange, Tineke de 173, 179
Laqueur, Walter 9
Laub, Leo 144
Lavon, Pinhas 13
Lazare, Lucien 227
Lazarus, Jacques 232–234
Leeuw, Channa de 20, 74, 94
Leeuw, Toos de 114
Lelièvre, Henri (Rik) 127, 131–134, 142, 144, 152, 158, 201, 258
Leuvenberg, Isaac 123–125
Leuvenberg, Jopie 123–125
Levin, Heinz 141
Levisson, company S 88
Levy, Marc 236
Levy, Ushi 77
Lichtenstein, Fritz 227
Lieme, Nehemia de 15
Ligt, Bart de 40f., 63, 67
Lind, Jakov 104, 142
Lindemans, Christiaan (King Kong) 222, 232
Linnewiel, Josef (Ad) 232f.
Litten, Manfred 90, 104–107
Litten-Serlui, Jansje (Shoshanna) 90, 104–107
Loeff, chairman of Montessori school 58
Loeff, Gerda 50, 55, 176f., 255
Lublin, Aron 218, 232
Lustig, Frits 164f., 170

Maanen, Gerda van 177, 179, 197, 205, 209, 261
Mann, Pim/Willem 33, 55, 176, 177, 179, 197, 205–207, 209, 252, 255, 256, 259.
Mechanicus, Philip 162f., 165, 169
Meierstein, Heinz 221
Meir, Golda 246
Middelburg, J. 104
Mogendorff, Hans 117, 140f., 164f., 193f.
Mok, Thea 192
Moll, de (company) 155
Montessori, Maria 47
Moore, Bob 250f.
Moses, Heinz 141

Mulder, 'Tante Kathi' 72
Mussolini, B. 126

Nehru, J. 41
Neumann, family 43
Nicolaas II 7
Nubert, Joep 187
Nussbaum, Gustel 74–76, 135
Nussbaum, Sophie 74–76, 135, 138f.

Okladek, Fritzi 232
Oliver, Pearl 250
Oliner, Samuel (Shmulek) 249f.
Oliveira, Jaap d' 97
Oroszlan, Carl (Carli) 168, 173, 208, 213
Osterer, Mosche 127f., 130, 132, 142, 227f.
Ouweneel, W.J. 34
Overduin, Leen 91, 245
Oversteegen, sisters 263

Pach, Ab 116
Paul, Manfred 73, 76f.
Penraat, Jaap 257f.
Perlmutter, Thea 200, 209
Pietersen, Marinus 77
Pinkhof, Juda 82
Pinkhof, Menachem 20f., 26f., 30, 66f., 69, 71f., 78f., 81, 87, 110, 116, 146, 148, 151f., 158, 172, 186, 188, 193, 197, 204, 206–209, 214, 229, 237, 239f., 261
Pohorylès, Henry 230f., 233f.
Polonski, Avraham 218f., 232
Posnanski, Walter 139
Poznanski, Renée 234
Presser, J. 1, 120, 166, 247
Prins, Saapke 97
Prins, Simon 97

Rehbein, Karl/Porel, Charles 230, 231–233, 237
Reichenberger, Mau 161
Reilinger, Kurt 109, 129f., 134, 140, 144–148, 150f., 155–158, 193f., 211, 217f., 220–224, 226, 229f., 232–235, 237, 240, 242, 245, 255, 258, 261–263

Rens, H. van 138
Reutlinger, Julius 179
Roitman, Jacques 219 f.
Roos, Antje (Ans) 94–96, 115 f., 226
Roos, Kees 94, 96
Rose, Leni 113
Roos, Riwka 177
Rose, Werner 113
Rosenbaum, David 171
Rosenbaum, Paul (Amo) 179, 206, 209
Rosenberg, Walter 154
Rothenberg, Klara 138
Rothmann, Rolf 155
Rothschild, Max 114
Rübner, Manfred 77, 190
Rudelsheim, family 195
Rudelsheim, Letty (Alida Jonker) 189–192, 194–196
Ruiter, couple 214, 261
Ruiter, Dirk de 186
Ruiter, Mrs. de (Maria van Ginkel) 98, 186, 187
Rümke, Philip 55, 67 f., 70, 80 f., 255

Sal, Meyer 223
Salomé, Chiel 134–136, 138, 147, 158, 170, 191 f., 196, 197, 254, 259
Salomé, Henk 134 f.
Samson, father 104, 160, 244
Samson, Manfred 104, 160 f., 184, 244
Sanders, Erich 221
Schabbing, Tinus 146 f., 156, 158, 171, 179, 197, 200, 212 f., 241, 254, 261
Schaft, Hanny 263
Schaik, Dirk van 104, 106 f., 115, 193
Schloss, Rolf 29, 145
Schönebaum, Gerd 113 f.
Schonfeld, (SD) 208
Schwalb, Nathan 119, 122, 124, 164
Sechestower, Heinz 127
Seeman, Hans 102 f.
Segboer, Tine 63, 76, 189, 206, 209 f., 253
Senechal, Derk 114, 117, 251
Sequerra, Samuel 227
Seyss-Inquart, A. 63, 253
Shazar, Zalman 246

Siegel, Paul 6, 13, 102, 110–113, 117, 161, 163, 165–168, 172, 174–177, 179, 190, 199 f., 225, 227 f., 263
Siesel, Frits 177
Sieskind, Lore 88, 92 f., 95 f., 98, 100, 109, 155 f., 187 f., 197, 199, 226, 229, 260, 263
Simon-Van Coevorden, Adina 121–125
Simon, Harald 190
Simon, Joachim (Shushu) 20, 21, 23, 29 f., 66, 71, 72, 74, 76, 79, 87, 93, 102, 119–123, 125–127, 129, 133 f., 144, 146, 148, 158, 186, 189, 194, 217 f., 258
Simons, Willi 145
Singer, Esther 171
Smedts, Mathieu 137
Smit, Jan 53, 68–70, 72, 77, 79, 81–84, 133, 134, 142, 146 f., 171, 176–178, 188, 197, 201, 203, 205 f., 209, 212–214, 241 f., 253, 255, 261, 263
Sonnenberg, Paul 73–76
Sperber, Gert 232–235
Sperber-Chlebowski, Anna 232–234
Spiegel, Lilo 157, 212, 230
Spier, Jo 77
Stein, Hans 184
Stertzenbach, Werner 167
Steyns, Dr. 204 f., 212
Stork, Greet 114 f., 141
Storm, Bram 38 f.
Streuvels, Stijn 198
Süsskind, Walter 197
Syrkin, Nahman 8

Tabenkin, Yitzhak 13
Tcherwinsky, Olga (Lydia) 230, 233
Tec, Nechama 251 f., 259
Teitelbaum, Israel 77
'Theo' 106, 142–144, 152, 258.
Tiefenbrunner, Israel 226
Tijn, Gertrude van 23 f.
Tischauer, Alice 28
Tischler, Ruth 138 f.
Todt, Fritz 153
Troelstra, P.J. 67
Trumpeldor, Josef 10

Turteltaub, Benno 77
Turteltaub, Max 77

Uffenheimer, Martin 171f., 175–179

Veen, Gerrit van der 96
Voûte, Hetty 257
Vries, Simon de 15
Vullinghs, Henri 135, 137–139, 251, 259

Wahrhaftig, Lotte 200
Walter, Kurt 173–175, 178f., 182, 188, 197
Ward, Justine 137
Waterman, Barend 119
Waterman, Chaja 64, 65, 239
Waterman, Elly 67, 72, 139, 213
Waterman, family 242
Waterman, Maurits 139
Waterman, Mirjam 30f., 57, 60, 64, 66f., 74, 79, 82f., 119, 170, 171, 188, 206, 208, 209, 214, 239, 253, 257, 261, 263
Weekers, M.C.E.H. 82f.
Weidner, Jean 258
Weill, Leo 232
Westerweel, Bart 41, 192

Westerweel, Joop (Johan Gerard, Victor) 32–39, 41–43, 45, 48f., 51–65, 68–70, 72, 74–76, 78f., 81–84, 100f., 115, 118f., 121–123, 126f., 130–135, 139, 142, 146f., 155, 157, 172, 176f., 188f., 192f., 197–206, 209–212, 214, 225, 233, 245–249, 251–255, 257–264
Westerweel, Leo 52
Westerweel, Martha 52, 247
Westerweel, Ruth 52, 115
Westerweel, Wil 45, 48f., 55, 59, 57, 63, 66, 78–80, 83, 114, 115, 119, 145, 188f., 191, 196, 197, 202, 206, 212, 242, 246f., 253, 255, 262f.
Windmüller, Emil 100, 155
Windmüller, Max (Cor) 147f., 158, 177, 199, 212, 217, 226, 229f., 232–237, 254, 261f.
Winter, Sophie (Fieke) de 193
Wolff, Paul 232–236
Wolff, Sam de 16, 93, 117
Wolterbeek Muller 36

Zurawel, Jacov 20

www.ingramcontent.com/pod-product-compliance
Lightning Source LLC
Chambersburg PA
CBHW031802220426
43662CB00007B/494